BASICS OF ADVANCE INORGANIC CHEMISTRY

BASICS OF ADVANCE INORGANIC CHEMISTRY

Prabir Pujhari

ANMOL PUBLICATIONS PVT. LTD.
NEW DELHI - 110 002 (INDIA)

ANMOL PUBLICATIONS PVT. LTD.
H.O.: 4374/4B, Ansari Road, Daryaganj,
New Delhi-110 002 (India)
Ph.: 23278000, 23261597
B.O.: No. 1015, Ist Main Road, BSK IIIrd Stage
IIIrd Phase, IIIrd Block,
Bangalore - 560 085 (India)
Visit us at: www.anmolpublications.com

Basics of Advance Inorganic Chemistry

First Published, 2008

ISBN 978-81-261-3569-1

PRINTED IN INDIA

Printed at Mehra Offset Press, Delhi.

Contents

Preface

This textbook is intended for the undergraduate or introductory graduate Inorganic chemistry course or sequence of courses, such as the one-semester junior-senior level Inorganic Chemistry course. It is intended to fill the gap between advanced inorganic textbooks that are encyclopedias of chemical reactions and those that are mainly textbooks of theoretical or physical inorganic chemistry. It covers virtually all topics suggested for the core inorganic requirement for a professional chemistry degree by the Indian Chemical Society Division of Inorganic Chemistry Ad-hoc Subcommittee.

This book is divided in "Atomic nuclei and structure and Simple Molecules in Chemistry and in our Environment," corresponds approximately to the first third of most advanced inorganic texts, but differs in that the chemical concepts and reactions. Bond theory, so that it is accessible to students who have had only general and organic chemistry courses. "Inorganic Substances and Materials: Theory and Applications," includes the remainder of the topics that are standard for an Inorganic course: symmetry, Molecular Orbital theory, materials chemistry, complex ion and inorganic reaction.

Author

Chapter 1

Introduction

Chemistry comprises two related but distinct activities:

1. The quest for an understanding of matter and material change,
2. The utilization of material change for human ends.

Ideally, the first activity provides the necessary know-how for the pursuit of the second, but in practice, the help it can give is only partial, and the second activity has to fall back on trial and error techniques in order to achieve its ends. This means that a good chemist is one who not only has a mastery of chemical theory, but also a good knowledge of chemical facts. With such a knowledge he can direct a trial-and error approach to practical problems in the most promising directions.

Organic chemistry is usually defined as the chemistry of compounds of carbon, inorganic chemistry being then the chemistry of all the other elements. This distinction is not a completely satisfactory one, however, since there are many compounds of carbon that are quite different from those studied by organic chemists (e.g. tungsten carbide, used for tipping cutting tools) and there are many compounds of other elements that are very similar to those studied under organic chemistry (e.g. the silicon analogues of the hydrocarbons). It is best, therefore, to think of inorganic chemistry as the chemistry of all the elements, with organic chemistry as being a more detailed study of certain important aspects of one of them - viz. the hydrocarbons and their derivatives.

Thinking of inorganic chemistry in this way brings together aspects of the chemistry of an element that would otherwise tend to become separated. For example, alcohols and ethers are usually dealt with under organic chemistry and are not thought of as being part of the chemistry of oxygen. Once they are, however, they can be set alongside the other compounds of oxygen, and a relationship immediately becomes apparent that might otherwise be lost, viz. that expressed by the formulae:

H–O–H   R–O–H   R–O–R′

Foundation Inorganic Chemistry covers, as the name suggests, the basic concepts of inorganic chemistry. If you have studied A level chemistry then all of what follows should be familiar to you.

STATES OF MATTER

Matter comes in three basic types, solid, liquid and gas (three phases of matter). In many cases it is possible to covert matter from one phase to another, perhaps by heating (we can melt, or fuse, a solid) or by cooling (the cooling of water to form ice for example).

Elements, Compounds and Mixtures

Elements are the basic building blocks of matter. If you examine the periodic table, you will see that there are about 110 elements, each given its own unique symbol. Of these hundred or so elements, about 90 exists naturally, either in their elemental (uncombined) form or in compounds where they have combined with one or more other elements. The remaining "non naturally occurring" elements have been produced artificially. Some of the elements will be familiar to you, oxygen, which is in the air around us, carbon which forms the "lead" in our pencils and diamonds and many of the metals which we see in common use such as iron and aluminium. Others may be less familiar, praseodymium, (Pr), promethium, (Pm) and protactinium, (Pa) for example. Some of these, more exotic elements have fierce sounding names but really we need not be concerned about these elements in the first year. We will, by and large, be concerned with the first 36 elements of the periodic table and you should make an effort to learn the names and chemical symbols of these 36 elements.When elements combine by chemical bonding, they form compounds. Compounds consist of two or more elements joined together by a chemical bond. Everyday examples include water (H_2O - two hydrogen atoms joined to an oxygen atom), table salt (NaCl - a sodium atom combining with a chlorine atom) and acetic acid (which is found in vinegar and involves carbon, oxygen and hydrogen bonding together to give CH_3COOH).When elements or compounds are mixed together and no chemical bonding occurs, we have a mixture, an example is the air we breathe which is a mixture (among other things) of nitrogen, oxygen, carbon dioxide and water vapour. The components of the mixture may be separated, so for example it is possible to liquefy air and obtain pure nitrogen and oxygen.

THE PERIODIC TABLE

Reference has already been made to the periodic table in these notes. The periodic table is a classification of the 100 or so elements. Each element has its own place in the table, its position determined by its atomic number. Hydrogen, with atomic number 1 is the first element. At first the arrangement of the elements within the periodic table may seem a little strange. Hydrogen, for example is often placed on it own with no particular position, sometimes it appears above the first column. Hydrogen and helium are found at the top of the table in a row of two elements, the next row (starting with lithium) contains eight elements (ending in neon, Ne). Then we have another row of eight followed by a row of eighteen and so on. The elements are the building blocks of matter, and to a certain extent this is true. However the elements themselves are built up from atoms. We could define an atom as the smallest particle of an element capable of displaying the properties of that element, although I am sure you will find alternative and probably better definitions elsewhere. This is not the end of the story however. Although atoms are very small (with diameters of about 10^{-10} m) they themselves are composed of sub-atomic particles. The three sub-atomic particles that concern us are the proton, the neutron and the electron. Both protons and neutrons reside in the nucleus of the atom (they are the nucleons). The nucleus can be thought of in terms of the core of the atom. The nucleus is very small, having a diameter of about 10^{-15}m. It is also very dense, both the proton and neutron have a mass of about 10^{-24}g, which although seems very small is confined in an extremely small volume. Given that the atom has a diameter of about 10^{-10}m and the diameter of the nucleus is about 10^{-15}m we can see that the vast majority of the atomic volume is outside the nucleus.

It is in this extra-nuclear region that is occupied by the electrons. The electron has a mass of about 10^{-28}g (i.e. 10000 lighter than the proton). We must be careful about how we describe the electron however since (and I hope this doesn't confuse you) it has the properties of a particle and of a wave. If you want to know more see my notes on atomic structure. These figures for mass are obviously not very convenient and so chemists have assigned different units to record atomic masses. The Atomic Mass Unit (amu) is used, 1 amu = 1.66 × 10^{-24}g. On this scale the mass of the proton and neutron is 1.0 amu and the electron is 1/1836 amu (or 5 × 10^{-4} amu).

Whilst the chemical properties of the elements are largely dictated by the number and arrangement of the electrons, it is the nucleons which dictate the mass of the atom (and hence the element). For reasons we will come to shortly, atoms are arrange not in terms of their mass but in terms of the number of protons in the nucleus, this is the atomic number (usually given the symbol Z).

CHARGES OF SUB-ATOMIC PARTICLES

The proton is a positively charged particle (+1) whilst the neutron is neutral and the electron is negatively charge (-1). In an uncharged atom, therefore, there will be an equal number of protons and electrons. Atoms may gain or lose electrons and in the process become ions. Atoms which lose electrons become cations (positively charged) and become anions when they gain electrons (negatively charged).

ISOTOPES

The chemical properties of the atom is by the number of electrons. Given that in a neutral atom, the number of electrons is equal to the number of protons, it follows that all atoms of a particular element will have the same number of protons. So all atoms of carbon, for instance, will have six protons. The same cannot be said for the number of neutrons. The number of neutrons in the nucleus of an element can vary. For example, the nucleus of a carbon atom can contain 6, 7 or 8 neutrons. Each of these is still a carbon atom, since each still has only six protons but the mass of each will depend upon the number of neutrons present. The mass number (A) of an element is given by the number of protons + the number of neutrons present. Hence we can have carbon-12 (with six protons + six neutrons), carbon-13 (six protons, seven neutrons) and carbon-14 (six protons and eight neutrons). These are isotopes of carbon. *Isotopes have the same value of Z but different values for A*. The mass of the element will then be determined by the relative proportions of the various isotopes present.

The atomic mass of chlorine is 35.45. Chlorine exists as a mixture of isotope Cl-35 and Cl-37. 75.53% of chlorine is Cl-35 and the remaining 24.47% as Cl-37. We can calculate the average mass of chlorine as (75.53% x 35) + (24.47% x 37) = 35.45

ELECTRONS

We have seen that the protons and neutrons are found in the nucleus and that electrons are "somewhere" outside the nucleus. Actually the term "somewhere" isn't as odd as at first it might seem, but that's another story. You will see elsewhere that electrons occupy energy levels outside the nucleus. We can classify them as "orbital". You should not, however, confuse this term with planetary orbits such as the earth around the sun. For the present, we will confine ourselves to recognising that certain energy levels exist within the atom.

Lower energy levels are close to the nucleus, higher energy levels further away. Electrons will occupy these energy levels starting by filling the lowest energy level possible. The first energy level, or shell can accommodate 2 electrons, the next level 8.

This starts to give us some idea of why the periodic table looks like it does. Energy levels are assigned a value, n. So n = 1 indicates the first shell. Each shell can contain 2n2 electrons. So when n=1 we have 2 electrons, n=2 we have 8 and when n=3 we have 18. Of course if you look at the periodic table this seems a little odd, since row 3 has only eight elements when we might expect 18. The explanation for this begins with something I said earlier about electrons entering energy levels at the lowest possible energy. And involves the make up of shells. Within each shell there are different types of orbitals. More detail on these are given in the atomic structure notes. Within the energy level n = 3, there are a group of orbitals labelled "3d", filling of these orbitals by electrons does not start until we get to the transition elements (starting atomic number 21 = Sc). Hence, the increase in the number of elements per row does not occur until we reach this point. When we reach scandium, the filling of the 3d orbitals begins and we have the series of 10 elements called the 3d transition metals.

SOME COMMENTS ON THE PERIODIC TABLE

The periodic table is divided into Groups and rows. The Groups are the columns, for example Group I which contains the alkali metals lithium, sodium, potassium, rubidium, caesium and francium. You may also see this labelled as Group 1A. The elements within each group will have similar properties, for example the Group I elements have a tendency to form compounds in which they are cations of +1 charge (NaCl being Na+Cl–).

On a more general note, the elements to the left of the periodic table are electropositive which means they tend to lose electrons when they form compounds.

- The elements to the right are electronegative, they tend to gain electrons.

- The elements in the centre (excluding the transition metals) tend to form compounds by sharing electrons.
- When an electropositive element combines with an electronegative element we get ionic bonding.
- When electropositive elements combine with other electropositive elements we get metallic bonding.
- When an electronegative element combines with another electronegative element we get covalent bonding.
- Ionic bonding involves the transfer of electron(s), covalent bonding involves the sharing of electrons.
- To the left of the periodic table we have the metals and to the right the non-metals.

Chapter 2

The Relative Importance of Different Elements and Compounds

All elements might be said to be equally important. The chemistry of each has to be understood in detail if the chemistry of matter as a whole is to be understood in detail. From the point of view of the world at large, however, the relative importance of different elements varies widely, and some elements are little more that chemical curiosities. In a university course of inorganic chemistry a balance has to be struck between a uniform treatment of all of the elements, and a greater concentration on the socially more important ones.

The question of which are the socially more important elements is to some extent a matter of opinion. A chemist working in the plastics industry would doubtless come up with a different list from one working in the metals industry, and someone working with agricultural chemicals would doubtless have different ideas from someone working on ceramics. Most people would agree, however, that among the most important elements are

a. The more abundant elements,

b. Elements that are consumed on a very large scale, and

c. Elements that are actively sought for despite very poor resources.

THE RELATIVE ABUNDANCE OF DIFFERENT ELEMENTS

The figures below give the relative abundance of different elements in the atmosphere, the hydrosphere, the lithosphere, and the biosphere. The figures are necessarily approximate, and slightly different values may be found in other places. The percentages are percentages by mass - the percentage numbers of atoms would of course be different. All abundances over 0.1% have been included. The figures reveal that some 99% of the matter in the world around us is made up of as few as 14 elements. Note, however, that many of the remaining elements in the biosphere, despite their low concentration, are essential to its healthy continuance.

Atmosphere		Hydrosphere	
Nitrogen	75.5%	Oxygen	85.8%
Oxygen	23.0%	Hydrogen	10.7%
Argon	1.4%	Chlorine	2.1%
		Sodium	1.1%
Total	**99.9%**	**Total**	**99.7%**
Lithosphere		**Biosphere**	
Oxygen	46.6%	Oxygen	52.4%
Silicon	27.7%	Carbon	39.4%
Aluminium	8.1%	Hydrogen	6.6%
Iron	5.0%	Nitrogen	0.5%
Calcium	3.6%	Calcium	0.4%
Sodium	2.8%	Potassium	0.2%
Potassium	2.6%		
Magnesium	2.1%		
Titanium	0.4%		
Total	**98.9%**	**Total**	**99.5%**

THE RELATIVE CONSUMPTION OF DIFFERENT ELEMENTS

The figures below give the annual world consumption of different elements, as elements and compounds, in millions of tons. The figures vary from year to year and are necessarily approximate. Siliceous materials and limestone used for building purposes have been excluded, and a lower limit has been set at 0.1 million tons.

	Millions of tons per year
Carbon	1000-10000
Iron, sodium	100-1000
Calcium, nitrogen, oxygen, potassium, sulfur	10-100
Aluminium, barium, chlorine, chromium, copper, fluorine, hydrogen, lead, magnesium, manganese, phosphorus, zinc	1-10
Boron, nickel, tin, titanium, zirconium	0.1-1

Of the 25 elements in the table, 10 are very plentiful, Of the remainder, the majority occur in relatively concentrated deposits, from which they or their compounds can be extracted fairly easily (e.g. hydrocarbons from petroleum). For copper, lead, zinc, nickel, and tin, demand is such that less concentrated deposits have to be worked, down to 0.5% in the case of copper and 0.2% in the case of tin.

ELEMENTS SOUGHT AFTER FROM VERY LOW-GRADE ORES

The following table gives the minimum metal content that an ore must have before it can be mined and smelted economically with present-day technology. Values are given up to a limit of 0.5%, and may be compared with the value for iron of 25%.

Platinum	0.0003%
Gold	0.001%
Silver	0.01%
Uranium	0.2%
Mercury	0.2%
Tin	0.2%
Molybdenum	0.3%
Copper	0.5%

CLASSIFICATION OF ELEMENTS INTO METALS AND NONMETALS

Types of Element

The classification of elements into metals and nonmetals, with the intermediate category of semimetal, is a fundamental one in inorganic chemistry.

Metallic elements are characterized by the following properties under ordinary conditions:

a. High lustre, high opacity over the visible spectrum;

b. High electrical and thermal conductivity.

Nonmetals are characterized by:

a. No lustre, high transparency over the visible spectrum;

b. Low electrical and thermal conductivity.

Semimetals are characterized by:

a. High opacity over the visible spectrum, little or no lustre;

b. Intermediate electrical conductivity.

"High opacity over the visible spectrum" is a way of saying that the substance absorbs light strongly over the whole range of the visible spectrum, or at least over the whole range beyond the red. If it is not lustrous, therefore, the substance appears black to the eye, or very dark brown.

"High transparency over the visible spectrum" is a way of saying that the substance absorbs light weakly over the greater part of the visible spectrum, if not over all of it. If no light is absorbed, the substance appears colourless or white. (Colourless substances look white when finely divided because light is scattered from the many crystal faces. Similarly, coloured substances become paler on grinding.) If some light is absorbed, the substance appears coloured, the colour being the complementary one to the colour of the light that is absorbed.

The complementary colours to the main spectral colours (red, orange, yellow, green, blue, and violet) are obtained by writing the colours in order round a circle. Those appearing opposite each other are complementary.

ELECTRICAL CONDUCTIVITY

$\kappa = 1/\rho$

The resistivity is defined by the equation

$R = \rho(l/A)$

where R is the resistance of a uniform conductor, l is its length, and A is its cross sectional area. The SI unit of ρ is ohm meter (Ω m) and its value is equal to the resistance that a meter cube of the substance offers to the passage of electricity from one parallel face to another. The SI unit of k is thus Ω^{-1} m^{-1}.On the basis of their ability to transmit an electric current, materials can be divided into four main classes:

Ordinary conductors, characterized by having conductivities greater than about 10^5 $\Omega^{-1}m^{-1}$ at room temperature, which fall with increase of temperature.

Semiconductors, characterized by having conductivities between about 10^{-7} and 10^5 $\Omega^{-1}m^{-1}$ at room temperature, which rise with increase of temperature, and are highly sensitive to traces of impurity.

Electrolytic conductors, characterized by the chemical decomposition that takes place at the points at which a direct current enters and leaves the material. (The words "electrolysis", "electrolyte", "electrolytic", come from the Greek *lyo,* to loosen.)

Insulators, characterized by conductivities of less than about 10^{-7} $\Omega^{-1}m^{-1}$ at room temperature, rising with increase in temperature.

In these terms, metals are conductors, semimetals are semiconductors, and nonmetals are insulators.

It is necessary to specify "under ordinary conditions" because, under extreme conditions, the character of a substance can change profoundly. Thus:

a. Gases at very low pressures become good conductors of electricity, as in a discharge tube at low pressure;
b. Metals at high temperatures vaporize, and lose their metallic properties completely (e.g. mercury boils at 357 ºC to give a colourless vapour, which is a poor conductor of electricity at ordinary pressures);
c. Semimetals and nonmetals at very high pressures become metallic in character (e.g. at 150 atmospheres iodine becomes a very good conductor of electricity).

A further characteristic of metals that is usually given is that, in the solid state, they are malleable (easily hammered into sheets) and ductile (easily drawn into wires). By contrast, nonmetals in the solid state are brittle and easily powdered. In practice, however, while the majority of elements that are metallic according to the above criteria are indeed malleable and ductile, there are a few that do not have this property, even when ultra-pure. (Slight traces of impurity can produce brittleness in metals. Thus, ordinary commercial tungsten is so brittle that it can only be worked with difficulty, whereas the ultra-pure metal can be cut with a hacksaw, turned, drawn, or extruded.) Since the exceptional elements include some whose metallicity has never bee questioned - e.g. manganese, cobalt, and zinc - it seems better *not* to make malleability and ductility a characteristic of metals, but rather a property possessed by most of them.

Zinc is a particularly interesting case, because it is malleable at some temperatures and brittle at others. Thus, it is brittle at room temperature, but softens above 100 ºC. At 205 ºC, however, it becomes brittle again, and can then be powdered in a mortar.

ASSIGNMENT OF ELEMENTS TO CLASSES

The clear-cut nonmetals are: hydrogen, nitrogen, oxygen, fluorine, sulfur, chlorine, bromine, and the inert or noble gases (helium, neon, argon, krypton, xenon).

The clear-cut semimetals are: boron, silicon, germanium, tellurium.

The clear-cut metals constitute all of the remaining elements, with the exception of the problem cases discussed below.

Carbon

Under ordinary conditions, carbon exists in two common forms, graphite and diamond, of which graphite is the more stable. Diamond is definitely nonmetallic, but graphite has properties on the borderline between a semimetal and a metal. Thus it is a shiny, black solid, readily separated into flakes, with a metallic conductivity in the plane of the flakes, and a semimetallic conductivity perpendicular to the plane. Other forms include black, petrol-soluble "fullerenes". When pure, these are insulators.

Phosphorus

This exists in a number of different forms: one white, several red, and several black. The white form can be made by condensing the vapour. It melts at 44 °C and boils at 280 °C. A red form is obtained when the white form is heated to just below its boiling point. A black form can be obtained by heating the white form under a very high pressure, or in the presence of a mercury catalyst. The yellow and red forms are insulators; the black forms are semiconductors. The most stable form under ordinary conditions is black; all forms, however, melt to a colourless liquid.

Arsenic

This can be obtained in several forms, ranging from a yellow one, which is similar to white phosphorus, to a grey one, which is a metal. The metallic form is the most stable under ordinary conditions; the yellow one is very unstable.

Antimony

This can also be obtained in several forms, ranging from yellow to metallic. The metallic form is again the most stable, with the yellow very unstable.

Selenium

This exists in three red forms, one black form, and one grey. The red and black forms are insulators; the grey is a semiconductor. The latter is the most stable.

Iodine

Iodine crystallizes in blackish-grey opaque scales, with an almost metallic lustre. In the plane of the scales it has the conductivity of a semiconductor; perpendicular to the plane it is an insulator.

VOLATILITIES OF METALS AND NONMETALS

The volatility of a substance is the ease with which the substance is vaporized. It is conveniently measured by the temperature at which the substance boils or sublimes at atmospheric pressure.

A broad generalisation concerning the relative volatilities of metals and nonmetals is the following:

Metals are generally involatile, i.e. they have boiling points in excess of 500 ºC. Only mercury has a boiling point below 500 ºC, this being at 357 ºC. Most metals have their boiling point somewhere between 1000 ºC and 4000 ºC. The highest value is for tungsten, at 5660 ºC.

Nonmetals vary from being very volatile to very involatile. All the definite nonmetals listed above have boiling points below 500 ºC, and most are gases at room temperature. The least volatile is sulfur, boiling at 445 ºC. If, however, we include carbon among the nonmetals, as surely we should, this lies at the opposite extreme of volatility, boiling at 4830 ºC.

CHEMICAL PROPERTIES OF METALS AND NONMETALS

There are many chemical differences between metals and nonmetals.

Some elements have very similar properties and are grouped into families.

These include:

Alkali metals: Li, Na, K, Rb, Cs
Alkaline earth metals: (Mg,) Ca, Sr, Ba
Halogens: F, Cl, Br, I
Inert or noble gases: He, Ne, Ar, Kr, Xe

"Alkali" is an Arabic word meaning "the ash" (ash from plants contains sodium and potassium carbonates).

"Halogen" means "producer-of (-gen) salt (Greek *hals*)".

Some chemists class magnesium as an alkaline earth metal as it is quite similar to calcium, often occurring with it in nature.

The gases He, Ne, Ar, Kr, and Xe were called "inert" when they were thought to be completely inactive. Their name was changed to "noble" after the xenon fluorides were discovered in 1962. This was because metals having a low activity (but nevertheless some activity) like gold and platinum are sometimes called "noble" as they are used for noble purposes. The change was inept. "Inert" need not imply complete inactivity, and "noble" is hardly apt (commoners breathe in these gases no less than nobles!).

CLASSIFICATION OF COMPOUNDS

Compounds can be classified according to the number of different elements that they contain: binary (two), ternary (three), etc.

LIMITING TYPES OF BINARY COMPOUND

Corresponding to the two limiting types of element, metal and nonmetal, there are three limiting types of binary compound:

a. *Metallic:* These are formed principally by the combination of a metal with a metal, and have the characteristics of a metal.

b. *Nonmetallic:* These are formed principally by the combination of a nonmetal with a nonmetal, and have the characteristics of a nonmetal.

c. *Salt-like:* These are formed principally by the combination of a metal with a nonmetal, and constitute a new class of material.

Metallic Compounds

An example is the compound Mg_2Sn. This is made by fusing together a mixture of magnesium and tin of the appropriate composition, and cooling the resulting melt. The product is homogeneous, as can be seen by cutting and polishing it, and examining the polished surface under a microscope. If more magnesium or tin is used than required by the formula Mg_2Sn, the product ceases to be homogeneous - crystals of Mg_2Sn can be seen embedded in a matrix of smaller crystals containing the excess of magnesium or tin. The compound Mg_2Sn is a bluish-white lustrous solid, having all the properties of a metal.

Metal-metal systems vary in character between the following extremes:

a. Those in which the two metals are able to form solid solutions with each other over the whole composition range from 100% of one to 100% of the other. An example of such a system is that between silver and gold.

b. Those in which the two metals are completely immiscible with each other, even in the liquid state. An example of such a pair of metals is iron and lead.

c. Those in which a number of compounds are formed.

Metallic compounds fall into one of two categories of compound:

a. *Stoicheiometric compounds:* These are compounds that obey the law of constant composition to very high accuracy, e.g. CH_4, NaCl, Mg_2Sn. They are also called "definite" compounds or "Daltonides".

   [The word "stoicheiometric" comes from "stoicheiometry", the name given to the measurement of the proportions of elements in compounds]

b. *Non-stoicheiometric compounds*: These have a variable composition within certain limits. They are also called "indefinite" compounds or "Berthollides".

Examples of non-stoicheiometric compounds are provided by the copper-tin system.

The three compounds referred to above have the following compositions at 300 ºC $CuZn_{0.85\text{-}0.95}$, $CuZn_{1.4\text{-}2.0}$, and $CuZn_{2.5\text{-}6.2}$.

The same compositions are obtained at room temperature if the solids are cooled fairly quickly. The first compound, however, is metastable at room temperature, and when cooled very slowly, decomposes at about 250 ºC into a mixture of the second compound and $CuZn_{0\text{-}0.5}$, a solid solution of zinc in copper (brass).

The formulae of compounds of this type are normally written in terms of some kind of idealized formula, using a *circa* ("round about") sign ~. The choice of idealized formula is based upon structural or theoretical considerations, and is not without arbitrariness. In the case of copper-zinc compounds, the idealized formulae are CuZn, Cu_5Zn_8, and $CuZn_3$, and the compounds are written ~CuZn, ~Cu_5Zn_8, and ~$CuZn_3$.

Notice that the first of these (CuZn) lies outside the composition range of the actual compound.

While metallic compounds are generally formed by the combination of a metal with a metal, they can sometimes be formed in the combination of a metal with a nonmetal. This occurs in circumstances where the influence of the metal can predominate over the influence of the nonmetal, namely

a. When the ratio of metal atoms to nonmetal atoms is relatively high,

b. When the nonmetallic character of the nonmetal is relatively weak, or

c. When both circumstances pertain.

An example of a compound of this type is silver subfluoride, Ag_2F, made by leaving metallic silver in contact with the normal fluoride, AgF, in the dark. It forms small crystals with a bronze reflex, and is a good conductor of electricity.Other examples of compounds of this type are the suboxides of rubidium and caesium, made by fusing the normal oxides, Rb_2O and Cs_2O, with the parent metals. These are lustrous and have good electrical conductivities, diminishing with increase of temperature. There are several for each metal, examples being Rb_9O_2, which is copper coloured, and Cs_7O, which is bronze.Another set of examples is provided by the lower oxides of titanium, made by heating titanium dioxide with the metal. These are lustrous and good conductors of electricity. They are non-stoicheiometric, their formulae being $\sim Ti_6O$, $\sim Ti_3O$, $\sim Ti_2O$, $Ti_{1.3\text{-}1.5}O$, and ~TiO.

To these may be added the nitrides of titanium, $\sim Ti_2N$ and ~TiN, which are also metallic in character.

Nonmetallic Compounds

These have the characteristic properties of nonmetals. Like the latter, they vary from being very volatile to very involatile. They are almost always stoicheiometric; only in the case of certain polymers does variable composition arise.

Examples of volatile nonmetallic compounds are very familiar: methane, ammonia, water. Examples of involatile compounds are silicon carbide (SiC) and silica (SiO_2). The former is manufactured by heating silica with graphite and is sold under the name carborundum. The commercial product is black, but when pure it is colourless. It melts at about 2700 ºC. Pure silica also forms colourless crystals, melting to a colourless liquid at about 1600 ºC and boiling at about 2400 ºC. Both compounds are insulators. Conductivity measurements have also been made on liquid silica, in which state it remains a poor conductor.Nonmetallic compounds are generally formed by the combination of a nonmetal with a nonmetal, but combinations of a metal and a nonmetal can also be nonmetallic if the influence of the nonmetal predominates. This occurs

a. When the ratio of metal atoms to nonmetal atoms is relatively low,

b. When the metallic character of the metal is relatively weak, or

c. When both circumstances pertain.

Good examples of nonmetallic compounds of metals are provided by the highest oxides of chromium and manganese. These have the formulae CrO_3 and Mn_2O_7, and are also similar to the highest oxides of sulfur and chlorine, SO_3 and Cl_2O_7. Thus CrO_3 is a red solid melting at 196 ºC, and Mn_2O_7 is a dark red oil. Further, electrical measurements on CrO_3 show this to be an insulator in both the solid and liquid state. Both compounds decompose below their boiling point, Mn_2O_7 explosively.

Salt-Like Compounds

These represent a new class, characterized by the following properties:

1. No lustre, high transparency over the visible spectrum,
2. Either
   a. Low electrical conductivity in the solid state, but high conductivity in the molten state and in aqueous solution, conduction being of the electrolytic type, or
   b. High electrolytic conductivity in both the solid and the liquid state.

They are brittle in the solid state, and generally have low volatilities.

An example of a compound of this type is of course salt itself (i.e. "common salt", sodium chloride). This forms colourless, brittle crystals, which melt to a colourless liquid at 801 ºC, and boil to a colourless vapour at 1413 ºC. In the solid, it has the electrical conductivity of an insulator, but the liquid has a conductivity of about $10^2 \Omega^{-1} m^{-1}$. When a direct current is passed through the melt, chlorine gas is evolved at the positive electrode (the anode) and metallic sodium is formed at the negative electrode (the cathode). This process is used industrially for the preparation of sodium, except that the temperature is lowered by the addition of another salt, e.g. calcium chloride, which depresses the freezing point of the sodium chloride.Sodium chloride also conducts electricity when dissolved in water. Strong solutions have a similar conductivity to the melt. A direct current again leads to chemical change at the electrodes, only this time hydrogen is produced at the cathode. This is to be expected, since sodium reacts with water to give hydrogen.Compounds that are good electrolytic conductors in the solid state are relatively rare. One example is silver iodide above about 150 ºC.The amounts of substance produced during electrolysis are given by Faraday's laws.

These May be Stated as Follows

First law: The amount of chemical change produced by a current is proportional to the quantity of electricity passed.

Second law: The quantity of electricity required to split up 1 mole of NaCl is the same as that required to split up 1 mole of KCl, or 1 mole of NaF, or ½ mole of

$MgCl_2$, or 1/3 mole of AlF_3, etc. This quantity is called the Faraday constant, *F*, and has the value 96500 coulombs per mole ($C\ mol^{-1}$).

Substances that give conducting solutions in water, but are themselves poor conductors in the liquid state, are not to be reckoned salt-like, but nonmetallic. For example, hydrogen chloride or hydrochloric acid gas, made by treating sodium chloride with concentrated sulfuric acid, dissolves in water to give a conducting solution. The gas itself, however, condenses at –85 ºC to a colourless liquid that is a very poor conductor of electricity. Hydrogen chloride is thus classed as a nonmetallic compound, and the electrical behaviour of its solution in water attributed to some change that takes place on dissolution. Evidence for such a change can be found in the high heat of solution of hydrogen chloride, as compared with that for sodium chloride:

	ΔH
HCl(g)	–74 kJ mol^{-1}
NaCl(c)	+4 kJ mol^{-1}

It is not easy to distinguish between a non-conducting salt-like compound and a nonmetallic compound. The only characteristic difference between them is in the conductivity of the melt, which can only be obtained at a high temperature. The following short-cuts, however, can be used fairly safely:

a. If the compound is soluble in water with a small heat of solution, and the solution conducts electricity, it is almost certainly salt-like.

b. If the compound is composed of similar elements to those of another compound whose character is known, and it has the elements in the same proportions, it probably belongs to the same class.

As an example of (ii), consider sodium oxide, Na_2O, and potassium oxide, K_2O. Since lithium oxide, Li_2O, is known to be a good electrolytic conductor in the liquid state, analogy suggests that Na_2O and K_2O should also be good electrolytic conductors in the liquid state, and therefore also salt-like.

Some metal-nonmetal pairs form a single compound (e.g. sodium and chlorine); others form more than one compound. For example:

- Sodium combines with a limited amount of oxygen to form sodium oxide (Na_2O) and with excess to form sodium peroxide (Na_2O_2). The latter reacts with water to give hydrogen peroxide (H_2O_2).
- Iron combines with chlorine to form iron dichloride ($FeCl_2$) and iron trichloride ($FeCl_3$), and with oxygen to form the oxides ~FeO, Fe_3O_4, and Fe_2O_3.
- Chromium and manganese form several compounds with oxygen, the main ones being Cr_2O_3, CrO_2, and CrO_3, and MnO, Mn_3O_4, Mn_2O_3, MnO_2, and Mn_2O_7. The highest of these were discussed above.

Pseudo-Binary Compounds

A great many compounds of ternary or higher order behave as if they were compounds of binary order, with a group of atoms acting as if it was a single atom. Such a group is called a "radical".

A classical example is provided by the cyanides, some of which are described below:

Hydrogen cyanide, HCN. This is manufactured by passing a mixture of methane and ammonia over a platinum catalyst at 1200 °C:

$$CH_4 + NH_3 \rightarrow HCN + 3H_2$$

It is a colourless liquid, boiling at 26 °C to a colourless gas. It is a poor conductor of electricity. Like most cyanides, it is extremely poisonous.

Sodium cyanide, NaCN. This is manufactured by absorbing gaseous hydrogen cyanide in aqueous sodium hydroxide or sodium carbonate. The compound is a colourless solid, melting at 564 ºC to a colourless liquid. The liquid conducts electricity with the formation of sodium at the cathode and a colourless gas at the anode. This gas is cyanogen, C_2N_2.

Potassium cyanide, KCN. This is similar to sodium cyanide.

Cyanogen, C_2N_2. This is a colourless gas, condensing at –21 ºC to a colourless liquid.

Cyanogen chloride, ClCN. This is made by passing chlorine into aqueous hydrogen cyanide:

$$HCN + Cl_2 \rightarrow HCl + ClCN$$

It is a colourless gas, condensing at 13 ºC to a colourless, mobile liquid.

From the above descriptions, it is clear that there is a close resemblance between cyanides and halides, both in composition and in character:

Composition		**Character**
Cyanide	*Chloride*	
HCN	HCl	Nonmetallic
NaCN	NaCl	Salt-like
KCN	KCl	Salt-like
ClCN	ClCN	Nonmetallic

This resemblance extends to the parent substances:

$(CN)_2$ Cl_2 Nonmetallic

So close is this resemblance that cyanogen is called a pseudo-halide. A second example is provided by ammonium compounds, e.g. ammonium chloride, NH_4Cl. These are very similar to alkali metal compounds. Indeed, ammonium compounds and potassium compounds often form mixed crystals with each other. Furthermore, electrolysis of ammonium chloride solution with a mercury cathode at 0 ºC gives a grey mass that is similar in character to sodium or potassium amalgam. Thus it reduces copper dichloride solution to copper, zinc chloride to zinc, etc. On warming it decomposes into ammonia and hydrogen in the proportions required by the equation

$2NH_4(\text{amalgam}) \rightarrow 2NH_3 + H_2$ The NH_4 radical is thus very much like an alkali metal atom in its behaviour.

The nature of pseudo-binary compounds

The nature of pseudo-binary compounds follows from our discussion of binary compounds. The bonding between radicals and radicals or radicals and atoms is the same as the bonding between atoms and atoms. For example :

a. The molecules HCN, ClCN, and $(CN)_2$ are held together with essentially covalent bonds: H–CN, Cl–CN, NC–CN
b. The compounds NaCN and KCN are ionic : $Na^+[CN]$, $K^+[CN]^-$
c. Ammonium chloride is ionic: $[NH_4]^+Cl^-$d.
d. Ammonium amalgam presumably contains: $[NH_4]^+e^-$

Bonding *within* radicals is of a covalent character. Thus the full bond formula of the HCN molecule is

H–CºN while that of the $[CN]^-$ ion is $^-$CºN Ions like $[CN]^-$ and $[NH_4]^+$ are called "molecular" ions. The square brackets indicate that the charge is somewhere on the radical as a whole, but they are often omitted: CN^- and NH_4^+. This should not be misinterpreted as implying, for example, that the charge on the cyanide ion is located on the nitrogen atom.

SOME IMPORTANT RADICALS

Formula of radical	Name as electropositive constituent of compounds	Name as electronegative constituent of compounds
Non-metal-like		
(a) Forming one covalent bond or a singly charged anion		
OH		hydroxide
NO_2	Nitryl	nitrite
NO_3		nitrate
CN	cyanogen	cyanide
SCN	thiocyanogen	thiocyanate
CH_3CO_2		acetate
CH_3 (Me)	methyl	methyl
C_6H_5 (Ph)	phenyl	phenyl
(b) Forming two covalent bonds or a doubly charged anion		
CO_3		carbonate
SO_3		sulfite
SO_4		sulfate
(c) Forming three covalent bonds or a triply charged anion		
PO_4		phosphate
Metal-like		
Forming a singly charged cation		
NH_4	ammonium	
NMe_4	tetramethylammonium	

a. All the nonmetal-like radicals form covalent bonds with sufficiently electronegative atoms or radicals. This includes nitrate and sulfate, e.g. in $MeNO_3$ and Me_2SO_4.

b. All the nonmetal-like radicals form anions with sufficiently electropositive elements. This includes methyl and phenyl, e.g. in Na^+Me^- and Na^+Ph^-.

c. The NMe_4 radical forms a similar amalgam to NH_4.

NOMENCLATURE OF RADICALS

The names given above are based on those recommended by IUPAC – the International Union of Pure and Applied Chemistry Though not very systematic, they reflect long established usage. The use of the endings –ite and –ate to indicate a lower and higher oxo-anion (e.g. nitrite, nitrate) parallels the use of –ous and –ic to indicate a lower and higher cation.

British schools teach different names as advised by the Association for Science Education (ASE):

IUPAC	ASE
Nitrite	Nitrate(III)
Nitrate	nitrate(V)
Sulfite	Sulfate(IV)
Sulfate	Sulfate(VI)
Acetate	Ethanoate

The ASE names are derived from IUPAC rules for naming less common ions. These names, however, are only used in British schools. Everywhere else in the world (in the chemical literature, chemical industry, other sciences, and other countries) IUPAC names are used.

NOMENCLATURE OF PSEUDO-BINARY COMPOUNDS

Pseudo-binary compounds are named in a similar way to binary compounds. In general, the name of the electropositive constituent is cited first, and that of the electronegative constituent second. Strictly speaking, this requires the incorporation of radicals in the electrochemical series, but in practice this is only done very roughly. Thus NO_3 and SO_4 are always regarded as being the electronegative constituent of a compound, even though in a few cases this is disputable, e.g. in FNO_3.

Examples of nomenclature:

CH_3Cl methyl chloride

$NaCH_3$ sodium methyl

$(CH_3)_2SO_4$ dimethyl sulfate

FNO_3 fluorine nitrate

(Note: some of these have alternative names, e.g. chloromethane.)

INTERMEDIATE TYPES OF BINARY COMPOUND

The three types of binary compound that we have been considering are what we have called "limiting types". In other words they represent extremes, and most binary compounds fall somewhere in between these extremes.

As can be seen from the following diagram, there are four intermediate types, corresponding to the one intermediate type between metallic and nonmetallic elements:

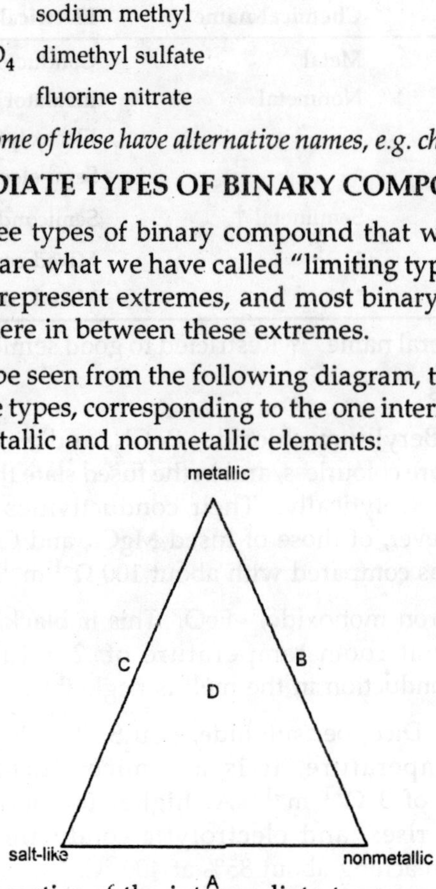

The properties of the intermediate types can be inferred from the properties of the limiting types.

Thus, *Type A:* High transparency, no lustre, weak electrolytic conductivity in fused state.

Type B: High opacity, possible some lustre, semiconducting in solid and liquid.

Type C: High opacity, possibly some lustre, both electrolytically conducting and semiconducting (i.e. with a direct current, chemical decomposition takes place, but less than the amount required by Faraday's laws).

Type D: Like C, but conductances weaker.

Type	Chemical name	Electrical name
Metallic	Metal	Conductor
Nonmetallic	Nonmetal	Insulator
Salt-like	*	Electrolyte
Type A	*	Semi-electrolyte
Type B	Semimetal †	Semiconductor
Type C	*	Mixed conductor
Type D	*	Mixed conductor

* No general name. † Restricted to good semiconductors.

Examples

Type A: Beryllium chloride, $BeCl_2$, and zinc chloride, $ZnCl_2$. These are colourless, and in the fused state they conduct electricity electrolytically. Their conductivities are only a fraction, however, of those of fused $MgCl_2$ and $CaCl_2$ (about 0×5 Ω^{-1} m^{-1}, as compared with about 100 Ω^{-1} m^{-1}).

Type B: Iron monoxide, ~FeO. This is black, and has a conductivity at room temperature of 2×10^3 Ω^{-1} m^{-1}. Electrolytic conduction in the melt is negligible.

Type C/D: Dicopper sulphide, ~Cu_2S. This is also black. At room temperature, it is a semiconductor, with a conductivity of 3 Ω^{-1} m^{-1}. At higher temperatures, the conductivity rises, and electrolytic conduction makes a contribution, reaching about 85% at 400 °C.

Volatile

The nature of compounds of this type can be arrived at by the following kind of argument:

a. Gases are readily compressed, whereas liquids and solids are not. This suggests that, in gases atoms are joined together in relatively small clusters, which are dispersed throughout the volume occupied by the gas.

b. Gases exert a pressure on the surface of a container. This suggests that the clusters of atoms are in continual, random motion, their bombardment of the walls creating the pressure we observe.

c. If the temperature of a gas is lowered at constant volume, its pressure drops (according approximately to the gas law, $P_1V_1/T_1 = P_2V_2/T_2$). This suggests that the lowering of the temperature has the effect of slowing down the motion of the clusters.

d. If the temperature of a gas is lowered far enough, the gas liquefies. If is cooled further, it solidifies. This suggests that there are forces between the clusters that are capable of holding the clusters together once their motion has been slowed down.

e. The distinction between volatile and involatile substances must therefore lie in the relative strengths of the forces between the clusters. In the case of volatile substances, the forces must be relatively weak.

f. This conclusion is supported by the following observations:

   i. The latent heats of vaporization of volatile substances are relatively low. For example:

	ΔH
Methane	10 kJ per mole of CH_4
Chlorine	20 kJ per mole of Cl_2
Salt	170 kJ per mole of NaCl
Iron	350 kJ per mole of Fe

   (The values of ΔH are for the vaporization of the liquid at the boiling point.)

   ii. Coloured volatile substances generally have the same colour in the liquid and the solid as they have in the gas, suggesting that the interactions between the clusters in the condensed state is relatively slight. For example, chlorine retains its yellow colour in the liquid and the solid: by contrast sodium is purple in the vapour and becomes metallic in the condensed state.

iii. Structure determinations show that, for volatile substances, the clusters that are present in the gaseous state are still present in the solid state, with relatively long distances between them. For example, electron diffraction of gaseous chlorine reveals the presence of diatomic clusters with an interatomic distance of 2.0 Å, while X-ray diffraction of solid chlorine reveals the presence of the same clusters, packed together in an orderly way, with the chlorine-chlorine distances *between* the clusters of 3.3 Å at their shortest.

vi. The presence of clusters in a volatile liquid, held together but moving about, is supported by the phenomenon of Brownian motion. When pollen grains are suspended in water and viewed with a microscope, they are seen to be undergoing a continual jerky motion, as if they are continually being bombarded by the clusters that make up the water.

In the above discussion "cluster" has been used to keep the argument fresh, and to avoid anticipating the conclusion. We can now substitute the technical word "molecule", from the Latin *molecula,* meaning "a little mass". The weak forces between molecules are called van der Waals' forces, after the Dutch physicist who invoked them to explain the deviation of real gases from the behaviour predicted by the *gas laws.*

ON THE CONSTITUTION OF MOLECULES

The formulae of the following molecules: H_2, Cl_2, HCl can be accounted for by supposing that hydrogen and chlorine atoms are capable of uniting with one other atom, almost as if each atom had a single hook on it, and could engage the hook of another atom to form a single connection ("bond") between them: H–H, Cl–Cl, H–Cl

Similarly the formulae of the following molecules: CH_4, CH_3Cl, CH_2Cl_2, $CHCl_3$, CCl_4 suggest that the carbon atom is capable of forming four such bonds:

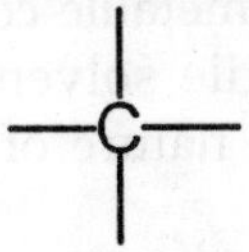

The formulae of a great many compounds can be rationalized in this kind of way, as is done routinely in organic chemistry. The number of bonds formed by an atom is called its valency. There is a good deal of evidence that the bonds around an atom are in definite directions in space. Thus, for example, compounds made up of molecules of the type CWXYZ can be resolved into optically active isomers, which would not be possible if the molecules were planar, but is consistent with their being tetrahedral in shape:

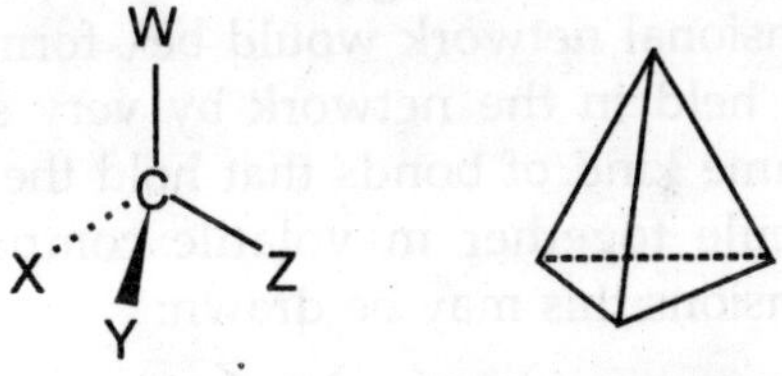

tetrahedron

This arrangement of four atoms around a carbon atom has been confirmed by numerous electron diffraction studies of carbon compounds in the gas phase and X-ray diffraction of carbon compounds in the solid state.

Involatile Nonmetallic Compounds

Some nonmetallic compounds are involatile because their molecules are very large (e.g. higher paraffins). Although the van der Waals interactions between atoms in different molecules are small (they are the same as the interactions between atoms of small molecules), there are so many atoms that the total interaction between molecules is large. Compounds of this type nevertheless dissolve in volatile solvents (e.g. higher paraffins in lower ones), because each molecule can interact with as many solvent molecules as it can get round it. These interactions add up to overcome the large interaction between molecules of the compound. Very large molecules are called "macromolecules".

Other involatile nonmetallic compounds are different in being insoluble in volatile solvents (e.g. diamond, silicon carbide, and silica). The nature of these compounds can be inferred as follows:

a. From the above discussion of volatile compounds, the molecules of these compounds in the vapour phase must interact very strongly in the liquid and solid phases.
b. In many volatile compounds, carbon atoms are known to bond to four other atoms in directions defined by a tetrahedron. Now if carbon atoms were to bond to each other in this kind of way, instead of a small molecule being produced, a continuous three-dimensional network would be\ formed, each atom being held in the network by very strong bonds - the same kind of bonds that hold the atoms of each molecule together in volatile compounds. In two dimensions this may be drawn:

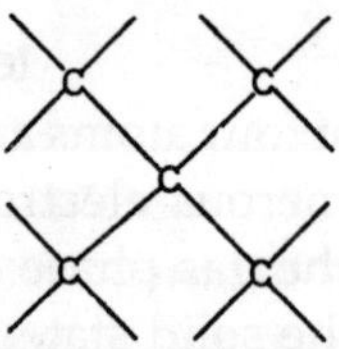

Such a structure would satisfy

a. And suggests itself for diamond.
b. A determination of the structure of diamond by X-ray crystallography shows that it indeed has this structure, with carbon-carbon bond lengths of 1.54 Å, as compared with 1.53 Å in C_2H_6.
c. In a similar way we can "guess" the structures of silicon carbide and silica. Silicon also forms volatile

compounds in which the silicon atom is surrounded by a tetrahedron of other atoms (e.g. simple molecules like SiH_4 and $SiCl_4$, and resclvable ones of the type SiWXYZ). Silicon carbide could thus have a similar structure to that of diamond:

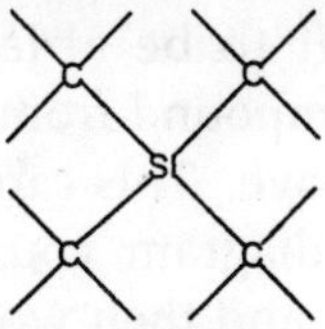

Further, since oxygen forms compounds in which its atom has two neighbours, silica could have a structure of the diamond type, only with an oxygen atom between every pair of silicon atoms:

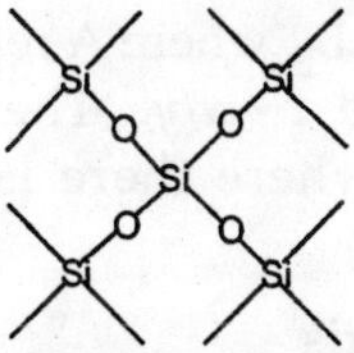

That silicon carbide and silica indeed have structures of this kind has been shown by X-ray crystallography. (Both compounds are polymorphic, but in each case the different forms have the same arrangement of nearest neighbours, and differ only in the relative orientation of more distant neighbours.)

Compounds of this type thus consist of giant assemblies of atoms tightly held together. Such assemblies of atoms are called "frameworks" and such compounds "nonmolecular". Liquefaction of compounds of this type involves the breaking of some of the bonds, for only in this way can atoms move positions. Clearly, however, the system will resist whole-scale breaking of bonds, and fluidity will rely on a process whereby the breaking of one bond is compensated for by the remaking of another bond somewhere else. Consequently the fluidity of such compounds in the liquid state is low: that is, their viscosity is high.

Vaporization of compounds of this type does require a break-up of the network of bonds, and the vapour effectively consists of fragments of the solid. In the case of diamond, these are C_2 molecules; for silica, they are SiO_2 molecules. Silicon carbide decomposes.

It is important to be able to work out the overall composition of a compound from diagrams of the kind given in the discussion above. This can be done by picking out the central atom in the diagram, counting the number of nearest neighbours to it (x), and then working out the share that the central atom has of these neighbours. This is done by counting the number of other atoms of the same type that the central atom has to share each neighbour with (y), from which the share the central atom has of each neighbour is $1/y$. The composition is then AB_z, where A represents the central atom, B its neighbour, and $z = x/y$. The method can readily be extended to cases where there is more than on kind of neighbour.

Salt-Like Compounds

These are composed of ions. The evidence for this is relatively simple:

a. A solution of salt in water conducts electricity: a solution of sugar does not. Something must be carrying the charge.

b. In dilute solution, the elevation of the boiling point of water or the depression of the freezing point is approximately twice as great for a certain number of moles of salt as it is for the same number of moles of sugar. Thus, instead of behaving like a solution containing NaCl molecules, it behaves as if twice as many particles are present. Between them these particles must be the carriers of the charge, and since salt itself is neutral, they must be either Na^{q+} and Cl^{q-} or Na^{q-} and Cl^{q+}.

c. When an electric current is passed through molten salt, sodium is formed at the negative electrode (the cathode) and chlorine at the positive (the anode). The corresponding products (hydrogen and chlorine) are obtained in aqueous solution. The particles must therefore be Na^{q+} and Cl^{q-}. Faraday called them "ions" from the Greek *ion* meaning "goer", "cations" being goers to the cathode and "anions" to the anode.

d. The value of q can be found by dividing the quantity of electricity required to deposit one mole of sodium, or produce the equivalent amount of hydrogen, F, by the Avogadro constant, L: $q = F/L$. This value was called the "electron" before the particle that bears this name was discovered. If we give it the symbol e, the sodium and chloride ions should strictly be written Na^{e+} and Cl^{e-} but the e is always dropped as being understood. The electron itself is written e^-, and its positive counterpart, the positron, e^+.

e. The idea that salt contains ions is further supported by the structure of the solid, as determined by X-ray diffraction. Instead of there being discrete molecules in the solid as there are in the case of chlorine, each sodium atom is surrounded by chlorine atoms, and each chlorine atom by sodium atoms, just as one would expect if they were oppositely charged ions.

f. Further support is provided by calculations of the energy of the solid based upon the idea that it is made up of fairly hard charged spheres. Good agreement with experiment is obtained.

g. Mass spectrometric analysis of sodium chloride vapour shows that, at temperatures a little above the boiling point, the vapour contains NaCl, Na_2Cl_2, and Na_3Cl_3 molecules, the proportion of Na_2Cl_2 and Na_3Cl_3 decreasing with increase of temperature. For NaCl, spectroscopic measurements give the sodium-chlorine distance (r) as 2.36 Å and the dipole moment (m) as $3.00'10^{-29}$ Coulomb meter. If m is conceived of as arising from a positive and negative charge of magnitude xe a distance r apart, we can write m =xer

From this and $e = 1.60'10^{-19}$ Coulomb we obtain $x = 0.8$. This corresponds to a charge distribution of $Na^{0.8+}Cl^{0.8-}$, which is reasonably close to that expected for an ion pair, Na^+Cl^-. Further, electron diffraction studies of Li_2Cl_2, which is formed in higher concentrations than Na_2Cl_2, indicate that it has the rhombus-shaped structure that we would expect it were made up of ions:

LI^+

Cl^- Cl^-

LI^+

The picture of sodium chloride that we arrive at is as follows:

a. In the solid, the compound is made up of ions, held together in a giant assembly by the attraction of opposite charges, each cation being immediately surrounded by anions, and each anion by cations.
b. In the liquid, the ions have enough kinetic energy to move about. Even so, the arrangement of the ions at any instant differs little from that in the solid, each cation being surrounded by a majority of anions and vice versa.
c. In the vapour, there are molecules, the molecules being small aggregates of ions. These readily condense into the liquid and solid since the electrostatic energy of the system is significantly higher when each ion has several neighbours of opposite charge than when it has a few.
d. At very high temperatures, the molecules dissociate into atoms, as when sodium chloride is introduced into a flame.

Metallic Compounds

The nature of these is more difficult to determine. However, we can get so far along the following line:

a. Metals conduct electricity without electrolysis taking place. The carriers of electricity cannot therefore be ions. They must either be some kind of charged entity that can travel through a conductor without affecting it or else a component of the atoms making up the conductor, which, by a process of successive displacement from one atom to another, can effectively be transferred from one end of the conductor to the other.

b. When an electric current is passed through a tube of gas at low pressure, rays can be seen to emerge from the negative electrode (the cathode). These rays are drawn towards a plate carrying a positive charge, and must therefore be carrying negative charge. They are likewise deflected by a magnet. From measurement of these deflections, the ratio of the charge to the mass can be determined for the particles making up the rays, and this is found to be the same, irrespective of the particular gas in the tube or the particular metal forming the cathode. This suggests that the particles are a common constituent of matter, and that they not only carry electricity through the gas, but also through the metal that supplies the electricity to the gas, the particles streaming without change from one into the other.

c. Measurement of the charge on these particles gives a value identical in magnitude with the charge on the sodium ion worked out above, viz. F/L. This gives us our terminology for the particle ("electron") and mode of conduction ("electronic"): it also gives us something else. It suggests that metal atoms may be thought of as being composed of metal ions and electrons, sodium atoms being Na^+ with one electron, magnesium atoms Mg^{2+} with two, and so on.

d. A simple model of a metal, therefore, is that of a collection of metal ions, held together by the appropriate number of electrons situated between them, the electrons being readily made to move by supplying further electrons at one point of the metal and taking electrons away at another. In the solid, the electrons hold the ions together in a rigid assembly: in the liquid the ions are free to move.

e. The constitutional formulae of metals can thus be written in the same way as the constitutional formulae of salts. The latter are properly written as exemplified by $(Na^+Cl^-)_n$, with *n* a very large number, but this is usually abbreviated to Na^+Cl^-, that is, to one unit of the complete assembly. In an analogous way, metallic sodium can be written Na^+e^-, metallic magnesium $Mg^{2+}(e^-)_2$, etc. A metallic compound like Cs_7O may similarly be written $(Cs^+)_7(e^-)_5O^{2-}$.

f. Support for these formulations is provided by the latent heats of vaporization of metals:

	ΔH
Sodium, Na^+e^-	90 kJ mol^{-1}
Magnesium, $Mg^{2+}(e^-)_2$	130 kJ mol^{-1}
Aluminium, $Al^{3+}(e^-)_3$	290 kJ mol^{-1}

These values are for the vaporization of the liquid at the boiling point into gaseous atoms. It can be seen that the strength of binding increases as the number of opposite charges increases.

The vapours of metals are, as we have seen, quite different from the solids and liquids. In general, metals vaporize into a mixture of atoms and diatomic molecules, the proportion of the latter diminishing with increase of temperature. Thus sodium metal just above the boiling point contains about 99% of sodium atoms and 1% of Na_2 molecules. On vaporization of a metal, therefore, the individual ions or pairs of ions become separated from each other, carrying their electrons with them. The mechanism for the conduction of electricity accordingly breaks down, and metallic properties disappear.

LIMITING TYPES OF CHEMICAL BOND

Corresponding to the three limiting types of binary compound there are three limiting types of chemical bond. These are named as follows:

Type of binary compound	**Type of chemical bonding**	
	Original name	**Modern name**
Metallic	Metallic	Metallic
Nonmetallic	Nonpolar	Covalent
Salt-like	Polar	Ionic

The character of these may be inferred from the nature of metallic, nonmetallic, and salt-like compounds.

Ionic bonding involves the transfer of electrons from metal atoms to nonmetal atoms and the binding together of the ions produced by the attraction of opposite charges.

Metallic bonding involves the binding together of metal ions by electrons instead of anions.

Covalent bonding involves some other kind of interaction between atoms, with neither the electrical polarization that is involved in ionic bonding nor the mobility of electrons that is possible with metallic bonding.

Between these limiting types are bonds of intermediate character, corresponding to the intermediate types of binary compound:

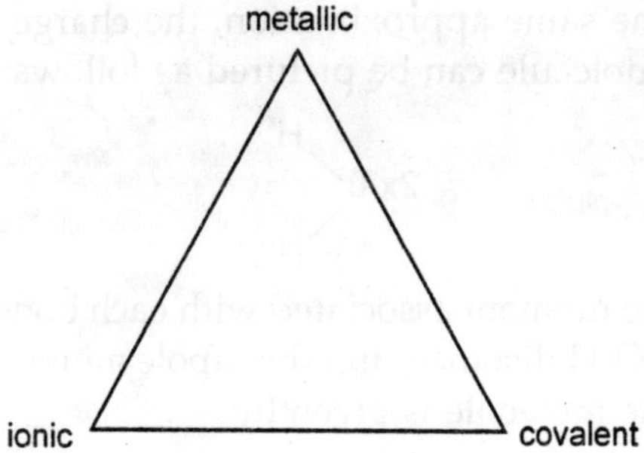

IONICITY

To determine the degree of ionic character (i) of a bond, we need to measure the charges on the atoms. This is in general difficult. Indeed, according to the quantum theory, this is something that cannot be done precisely even in principle.

However, we saw earlier that the charges on the atoms in a diatomic molecule (AB) can be approximately determined from its dipole moment (m). If there is transfer of charge from atom A to atom B to give $A^{x+}B^{x-}$, the value of x is approximately given by

$$\mu = xer$$

where r is the internuclear distance. The value is approximate because there may be displacement of charge *within* the atoms contributing to μ.

In the table below, values of x are given for hydrogen halide molecules, as calculated from μ and r. These give the values of i shown in the last column.

	r/Å	$\mu/10^{-30}$ C m	x	i
HF	0.917	6.07	0.41	41%
HCl	1.275	3.60	0.18	18%
HBr	1.41	2.74	0.12	12%
HI	1.600	1.47	0.06	6%

This method can be extended to other molecules. For example, to the same approximation, the charge distribution in the water molecule can be pictured as follows:

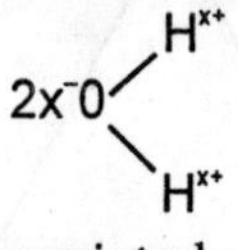

The dipole moment associated with each bond is then xer, where r is the O-H distance, and the dipole moment associated with the whole molecule is given by

$$\mu = 2xer \cos \tfrac{1}{2}\theta$$

Where q is the H-O-H angle. From the observed values of r (0.957 Å), θ (104.5°), and μ (6.17×10^{-30} C m), x is calculated to be 0.33. The charge distribution is thus $O^{0.66-}(H^{0.33+})_2$, and the ionic character of each bond 33%. The considerable polarity explains many of the unusual properties of water.

This method rests on the assumption that there is little displacement of charge within the atoms concerned. Quantum-mechanical calculations support this assumption for the above molecules, giving similar values of x:

HF ~ 0.5   HCl ~ 0.2   HBr ~ 0.1   HI ~ 0.05 H_2O ~ 0.4

However, the assumption does not hold for all molecules.

For bonds in solids, assessing the degree of ionic character is more difficult. That this can vary, however, is clear enough. For example, cadmium chloride ($CdCl_2$) has a less ionic kind of structure than sodium chloride. In the latter, as we have seen, cations are surrounded by anions and anions by cations. In $CdCl_2$, however, while the cations are surrounded by anions the anions only have cations on one side, to give a layer structure. Mercury dichloride ($HgCl_2$) has an even more covalent type of structure, in which there are discrete, linear $HgCl_2$ units, similar to the molecules in the vapour.

METALLICITY

The degree of metallic character of a bond in a binary compound is best measured by the electrical conductivity of the compound. Strictly speaking, the property is not of an isolated bond (as in Na_2) but of bonds in concert.

TYPES OF FORMULA

Empirical formulae

The emplrical formula of a compound is the simplest possible formula indicating the relative numbers of different atoms in it. The formula is determined from the composition of the compound.

Molecular Formulae

The molecular formula of a compound indicates the total numbers of different atoms in a molecule of it. For molecular compounds, the molecular formula is determined from the empirical formula and the molecular mass of the compound. For nonmolecular compounds, it is set at the empirical formula multiplied by infinity.

In practice, a mixture of empirical and molecular formulae are used – NaCl alongside C_2H_6. Even hybrid formulae are used, e.g. Na_2O_2 (as this compound contains O_2^{2-} ions). Ideally, the different formulae should be distinguished, e.g. with brackets: (NaCl) $[C_2H_6]$ $(Na_2[O_2])$

Structural formulae

The structural formula of a compound indicates the way in which the atoms are united in the molecule. There are two main types:

1. *Constitutional formulae,*
2. *Descriptive formulae.*

1. Constitutional formulae:
   a. *Covalent compounds: bond formulae* These are determined from the valencies of the atoms and such structural information as is required.
   Example:

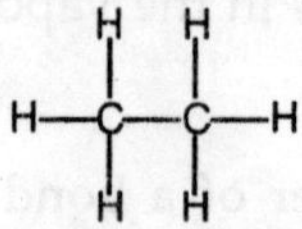

   b. *Ionic compounds: ion formulae* These are determined from the charges on the ions.
   Example: Na^+Cl^-

2. Descriptive formulae: These do not attempt to give any indication of the bonding.
   a. Single stroke formulae: These are formulae in which bonded atoms are connected up with a single stroke, with no indication of bond number or polarity. They are usually used for compounds in which the bonds have an intermediate bond number or polarity. An example is the following representation of one layer of graphite:
   Example:

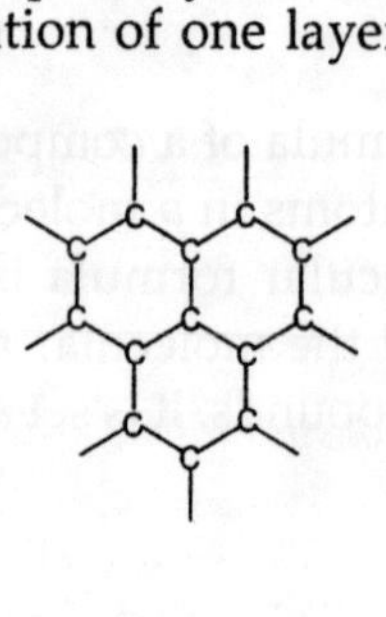

This kind of representation is also used in referring to bond distances in a compound, as in the statement "the C–C bond length in graphite is 1.42 Å". Care needs to be taken not to confuse formulae of this kind with bond formulae.

b. *Descriptive formulae of crystals:* If the crystal structure of a nonmolecular compound is known, the arrangement of atoms can be represented in the form $[AB_{x/y}]_{\infty}$, or more generally $[AB_{x/y}B'_{x'/y'}\cdots]_{\infty}$, where x is the number of nearest neighbours B to A, and y is the number of nearest neighbours A to B – $1/y$ being the share atom A has of each neighbouring atom B .

Example: $[NaCl_{6/6}]_{\infty}$

This indicates that in crystalline sodium chloride each sodium atom (ion) is surrounded by six chlorine atoms (ions), each of which is shared between six sodium atoms (ions), giving an empirical formula NaCl. Crystallographers call the number of nearest neighbours the "coordination number".

Geometrical Formulae

The geometrical formula of a compound represents the geometrical arrangement of the atoms in the molecule. It is determined from the structure of the compound. To represent the geometry properly really requires a three-dimensional model, but we have to try to do this in two dimensions.

Example:

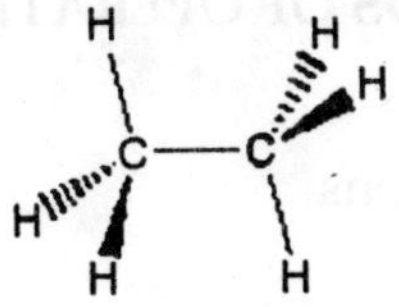

Chapter 3

Molecular Symmetry

SYMMETRY AND GROUP THEORY

Cataloging the symmetry of molecules is very useful.Group Theory is a mathematical method by which aspects of a molecules symmetry can be determined.

The symmetry of a molecule reveals information about its properties (i.e., structure, spectra, polarity, chirality, etc...)Clearly, the symmetry of the linear molecule A-B-A is different from A-A-B. In A-B-A the A-B bonds are equivalent, but in A-A-B they are not.However, important aspects of the symmetry of H_2O and CF_2Cl_2 are the same. This is not obvious without Group Theory. Symmetry Operations/ElementsA molecule or object is said to possess a particular operation if that operation when applied leaves the molecule unchanged. Each operation is performed relative to a point, line, or plane - called a symmetry element.

THERE ARE 5 KINDS OF OPERATIONS

1. Identity
2. n-Fold Rotations
3. Reflection
4. Inversion
5. Improper n-Fold Rotation

1. Identity is indicated as E does nothing, has no effect all molecules/objects possess the identity operation,

i.e., posses E. E has the same importance as the number 1 does in multiplication (E is needed in order to define inverses).

2. n-Fold Rotations: Cn, where n is an integer rotation by 360°/n about a particular axis defined as the n fold rotation axis.

   C_2 = 180° rotation, C_3 = 120° rotation, C_4 = 90° rotation, C_5 = 72° rotation, C_6 = 60° rotation, etc. Rotation of H_2O about the axis shown by 180° (C_2) gives the same molecule back. Therefore H_2O possess the C_2 symmetry element.

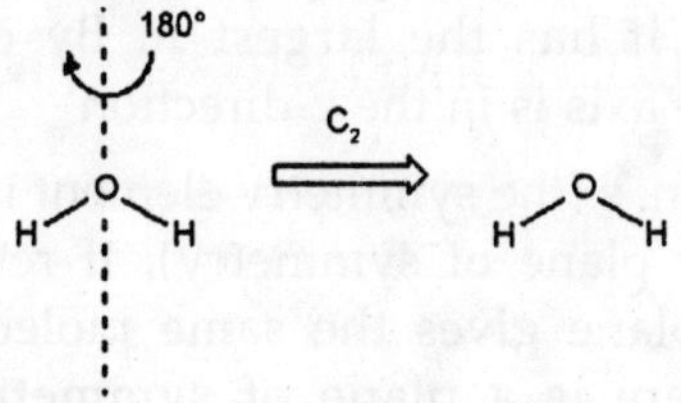

However, rotation by 90° about the same axis does not give back the identical molecule. Therefore H_2O does NOT possess a C_4 symmetry axis.

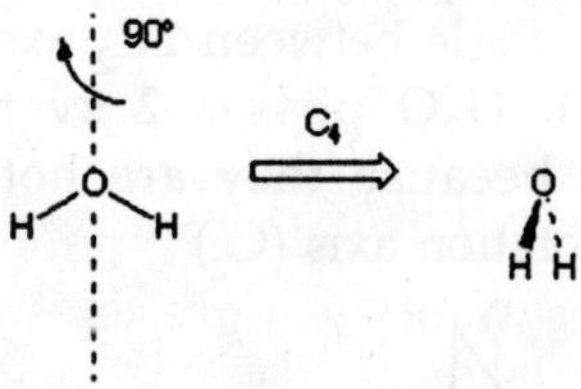

BF3 posses a C_3 rotation axis of symmetry. (Both directions of rotation must be considered)

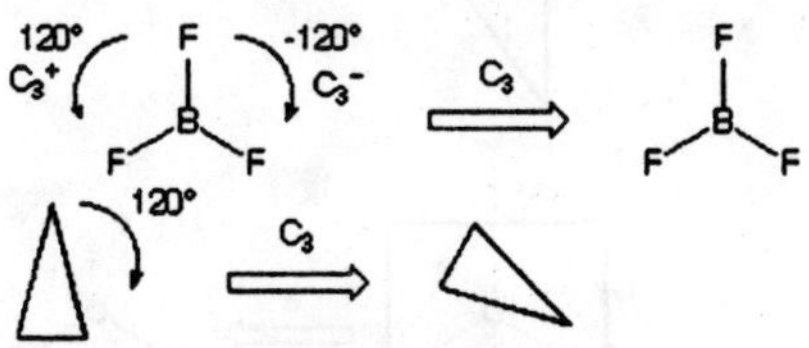

This triangle does not posses a C_3 rotation axis o symmetry.

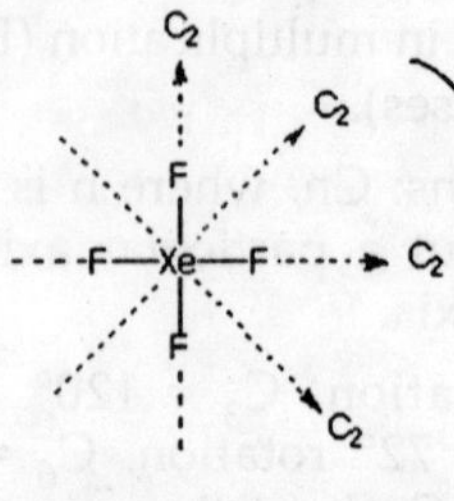

XeF_4 is square planar.

It has four different C_2 axes. It also has a C_4 axis coming out of the page called the principle axis because it has the largest n. By convention, the principle axis is in the z-direction

3. Reflection: s (the symmetry element is called a mirror plane or plane of symmetry). If reflection about a mirror plane gives the same molecule/object back than there is a plane of symmetry (s). If plane contains the principle rotation axis (i.e., parallel), it is a vertical plane (sv). If plane is perpendicular to the principle rotation axis, it is a horizontal plane (sh). If plane is parallel to the principle rotation axis, but bisects angle between $2C_2$ axes, it is a diagonal plane (sd). H_2O posses 2 sv mirror planes of symmetry because they are both parallel to the principle rotation axis (C_2)

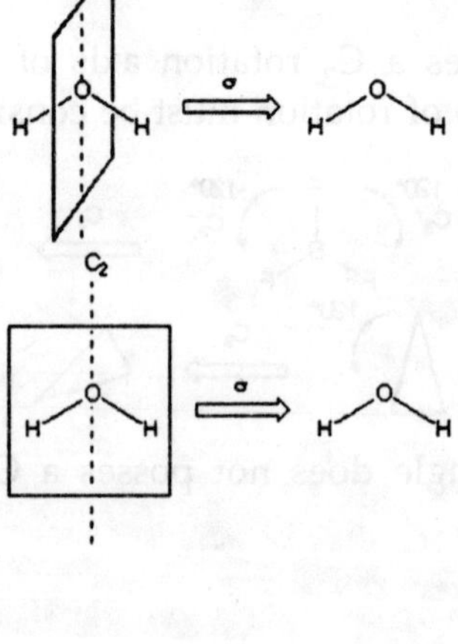

XeF_4 has two planes of symmetry parallel to the principle rotation axis: sv

XeF_4 has two planes of symmetry parallel to the principle rotation axis and bisecting the angle between $2C_2$ axes : sd

XeF_4 has one plane of symmetry perpendicular to the principle rotation axis: sh

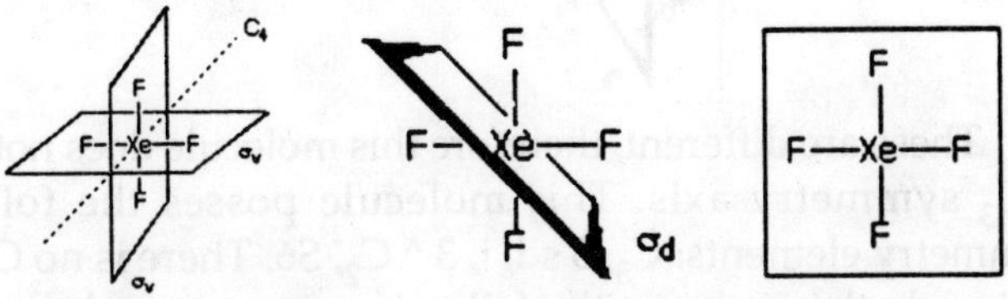

4. Inversion: i (the element that corresponds to this operation is a center of symmetry or inversion center) The operation is to move every atom in the molecule in a straight line through the inversion center to the opposite side of the molecule.

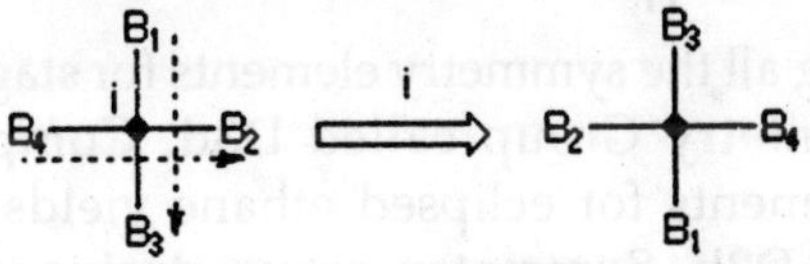

Therefore XeF_4 posses an inversion center at the Xe atom.

5. Improper Rotations: S_n n-fold rotation followed by reflection through mirror plane perpendicular to rotation axis

   Note: n is always 3 or larger because S_1 = S and S_2 = i.

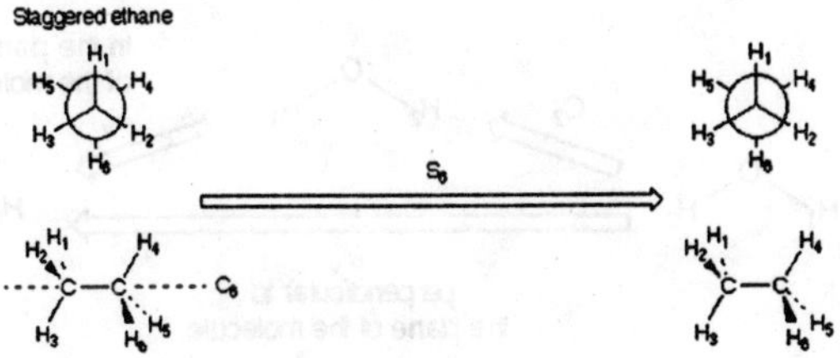

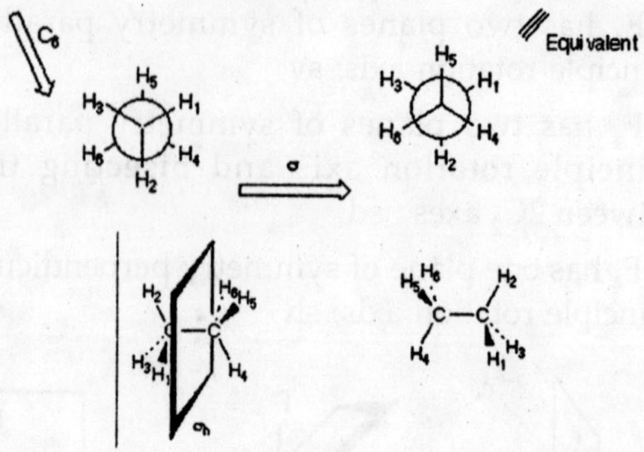

These are different, therefore this molecule does not posses a C_3 symmetry axis. This molecule posses the following symmetry elements: C_3, 3 sd, i, 3 ^ C_2, S6. There is no C_3 or sh. Eclipsed ethane posses the following symmetry elements: C_3, 3 sv, 3 ^ C_2, S_3, sh. There is no S6 or i.

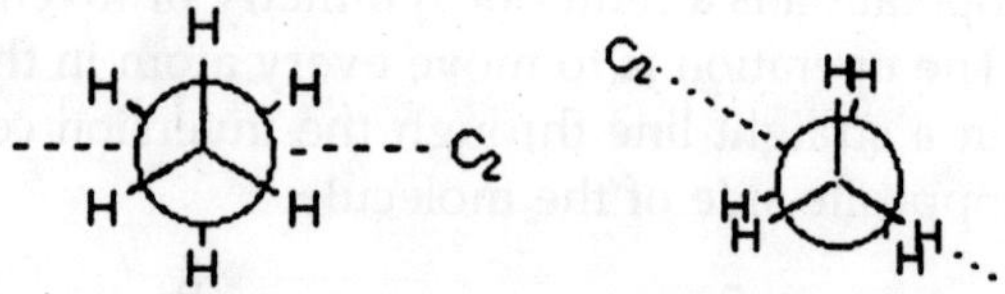

Compiling all the symmetry elements for staggered ethane yields a Symmetry Group called D3d. Compiling all the symmetry elements for eclipsed ethane yields a Symmetry Group called D3h. Symmetry group designations will be discussed in detail shortly

To be a group several conditions must be met:

1. Any result of two or more operations must produce the same result as application of one operation within the group. i.e., the group multiplication table must be closed. Consider H2O which has E, C2 and 2 sv's.

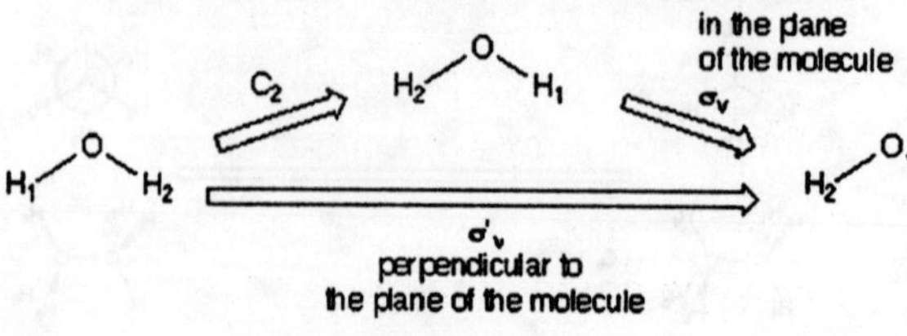

i.e., $\hat{C}_2 \hat{\sigma} \equiv \hat{\sigma}v$

of course $\hat{C}_2 \hat{C}_2 \equiv \hat{E}$ etc...

The group multiplication table obtained is therefore:

	E	C2	sv	s′v
E	E	C2	sv	s′v
C2	C2	E	s′v	sv
sv	sv	s′v	E	C2
s′v	s′v	sv	C2	E

Note: the table is closed, i.e., the results of two operations is an operation in the group.

2. Must have an identity ($\hat{E}$)
3. All elements must have an inverse i.e., for a given operation ($\hat{A}$) there must exist an operation ($\hat{B}$) such that $\hat{A}\hat{B} = \hat{E}$.

Classification of the Symmetry of Molecules

Certain symmetry operations can be present simultaneously, while others cannot.

There are certain combinations of symmetry operations which can occur together.

Symmetry Groups combine symmetry operations that can occur together.

Symmetry groups contain elements and there mathematical operations.

For example, one of the symmetry element of H_2O is a C_2-axis. The corresponding operation is rotation of the molecule by 180° about an axis.

- Point Groups
- Low Symmetry Groups

C1: only E

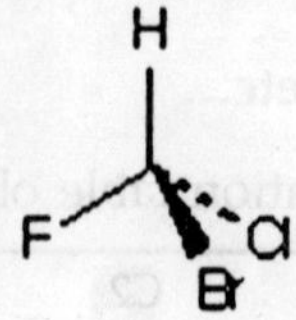

Cs: E and s only

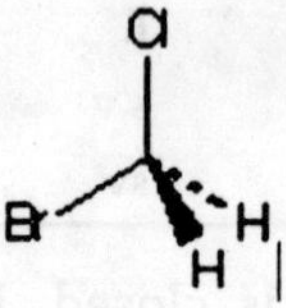

Ci: E and i only

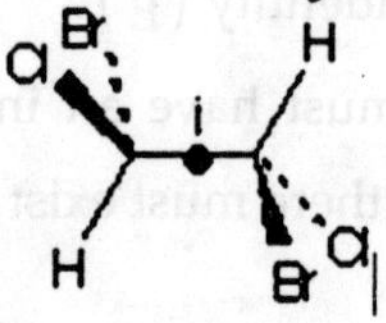

Cn, Cnv, Cnh Groups

Cn: E and Cn only

C_2:

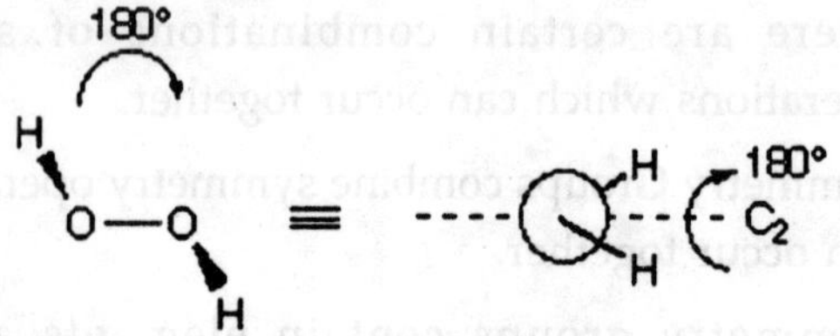

C_3:

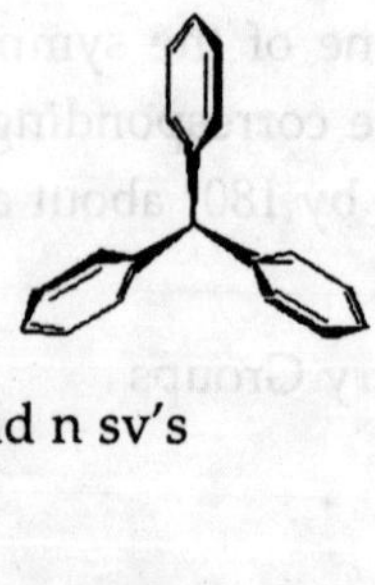

Cnv: E and Cn and n sv's

C2v: E, C_2, 2 sv H_2O

C3v: E, C_3, 3 sv NH_3

C• v: E, C• , • sv HF, HCN

Cnh: E and Cn and sh (and others as well)

C_2h: E, C_2, sh, i

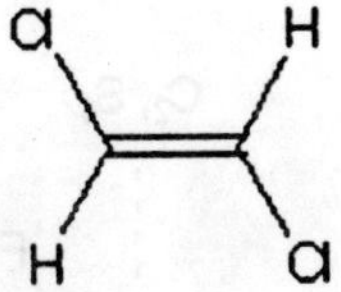

D_n, Dnv, Dnh Groups

D_n: E, Cn, n C_2 axes ^ to Cn

D_3: E, C_3, 3 ^ C_2

$[Co(en)_3]^{3+}$

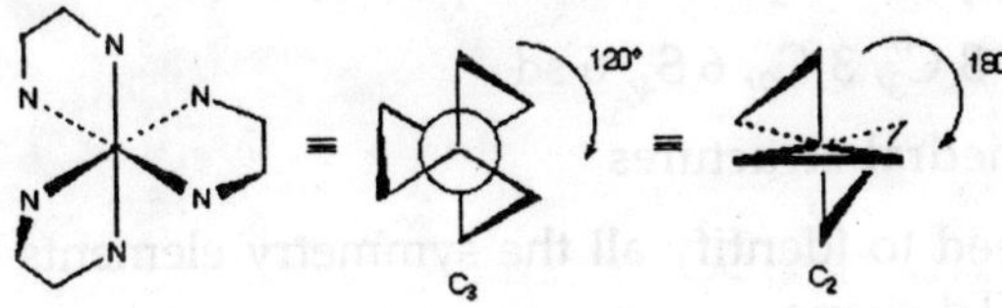

Dnh: E, C_n, n C_2 axes ^ to C_n, sh

D3h: E, C_3, 3 ^ C_2, sh

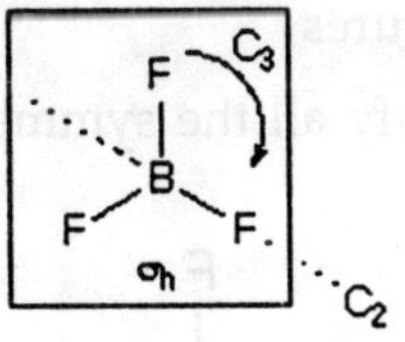

D6h: E, C_6, 6 ^ C_2, sh

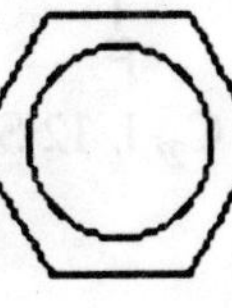

$O{=}C{=}OH_2$

Dnd: E, C_n, n C_2 axes ^ to Cn, D3d: E, C_3, 3 ^ C_2, 3 sd

staggered ethaneS_n Group

S_{2n}: E, Cn, S_{2n} (no mirror planes)

S_4, S_6, S_8, etc. (Note: never S_3, S_5, etc.)

S_4: E, C_2, S_4

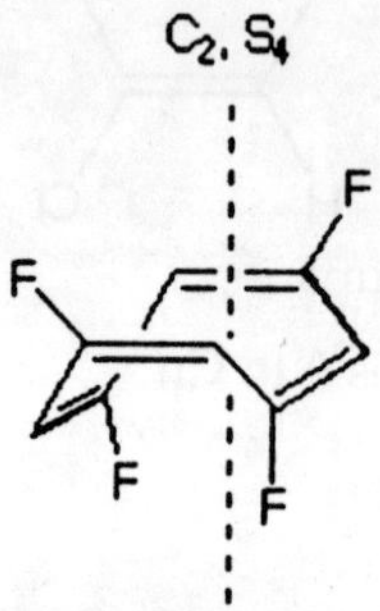

High Symmetry Cubic Groups, Td, Oh, Ih

Td: E, 8 C_3, 3 C_2, 6 S_4, 6 sd

Tetrahedral structures

No need to identify all the symmetry elements - simply recognize Td shape.

Methane, CH_4

Oh: E, 8 C_3, 6 C_2, 6 C_4, i, 6 S_4, 8 S_6, 3 sh, 6 sd

Octahedral structures

No need to identify all the symmetry elements - simply recognize Oh shape.

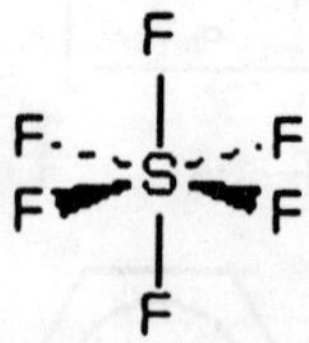

Ih: E, 12 C_5, 20 C_3, 15 C_2, i, 12 S_{10}, 20 S6, 15 s

Icosahedron

Other rare high symmetry groups are T, Th, O, and I

Molecular structures

Linear	O=C=O
Planar triangle	SO_3
Tetrahedral	CH_4
Trigonal bipyramidal	PCl_5
Octahedral	SF_6
Pentagonal bipyramidal	IF_7
Square antiprism	TaF_8^{3-}

Symmetry Operations: An action such as reflection, rotation, ...

Symmetry Elements: Associated with each symmetry operation, a plane, a line , a point, ...

A symmetry element is an axis (line), plane, or point about which symmetry operations are performed.

Aspects of Symmetry and Point Groups

Symmetry operations: Proper rotation, C_n

Reflection, σ

Inversion, i

Improper rotation, S_n

Proper rotation, C_n simple rotation about an axis passing through the molecule by an angle of 360°/n. This operation is called a proper rotation and is symbolized by C_n

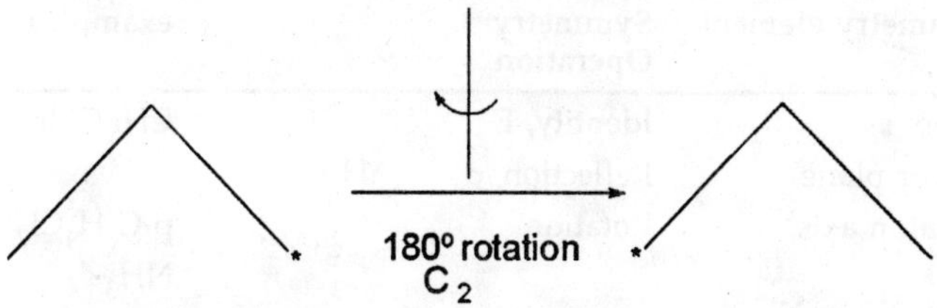

For a H_2O molecule, please locate the C_2 axis

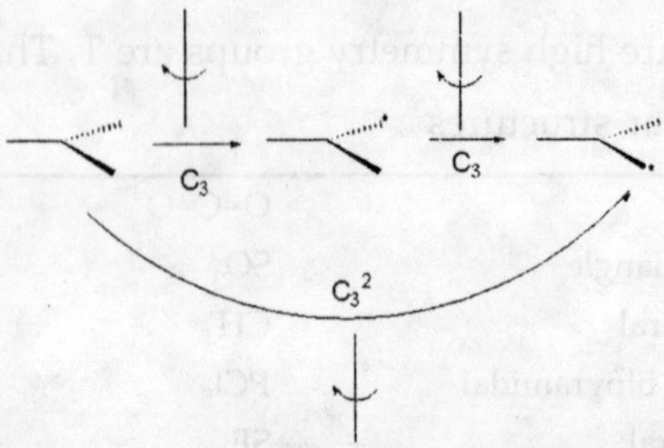

For a BF_3 molecule, please locate the C_3 axis

Reflection (mirror) plane, s Reflection of all atoms through a plane Please locate the two vertical reflection planes in H_2O

Inversion (Center of Inversion), I Reflection of all atoms through a point Please locate the center of inversion in SF_6, XeF_4, etc.

Improper rotation, S_n The combination, in either order, of rotating the molecule about an axis passing through it by $2\pi/n$ and reflecting all atoms through a plane that is perpendicular to the axis of rotation

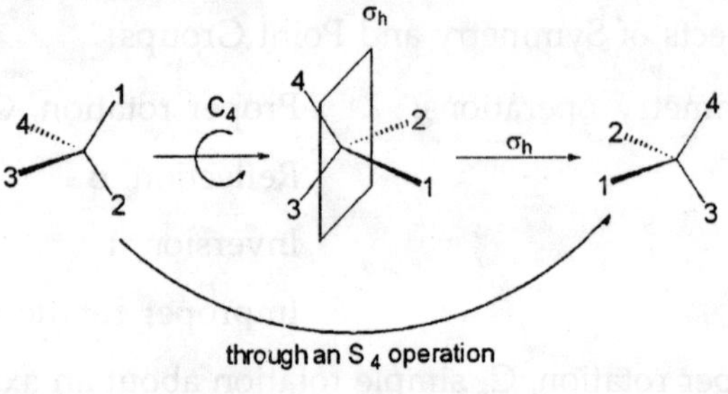

An S_4 improper rotation axis

How many S_4 rotation axes for a CH_4 molecule?

Symmetry element	Symmetry Operation		examples
None	Identity, E		CHFClBr
Mirror plane	Reflection, σ	H_2O	
Rotation axis	Rotation	C_2	p-$C_6H_4Cl_2$
		C_3	NH_3
		C_4	$[PtCl_4]^{2-}$

		C_5	$C_5H_5^-$
		C_6	benzene
Inversion center	Inversion, i		Ferrocene (staggered)
Improper axis	Rotation -reflection	S_4	CH_4
		S_6	Ethane (staggered)
		S_{10}	Ferrocene (staggered)

Some important points

1. Do not confuse an inversion operation with a two fold rotation

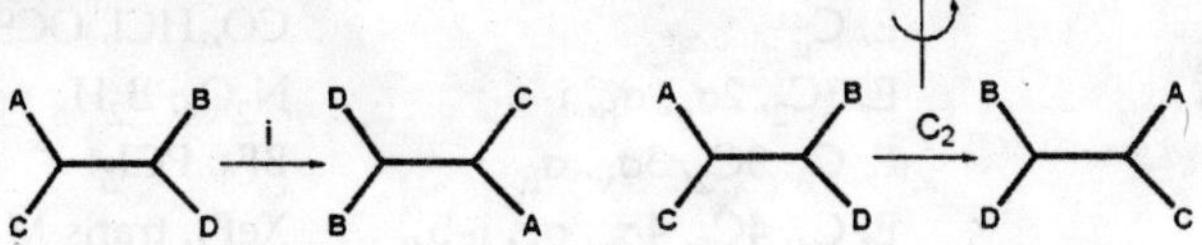

2. S_1 operation is equivalent to the mirror reflection σ
3. S_2 is equivalent to a center of inversion

Summary of important symmetry operations and symmetry elements

Symmetry element	Symmetry operation	Symbol
	Identity	E
n-Fold symmetry axis	Rotation by $2\pi/n$	C_n
Mirror plane	Reflection	σ
Center of inversion	Inversion	I
n-Fold axis of improper rotation	Rotation by $2\pi/n$ followed by reflection perpendicular to rotation axis	S_n

One molecule may have more than one symmetry elements

NH_3 H_2O

SO_2Cl_2 SF_6

To summarize the collection of symmetry elements a molecule has, we introduce the concept of point group.

POINT GROUPS (OR SYMMETRY GROUPS)

The complete set of symmetry operations that can be performed on a molecule is called the symmetry group for the molecule

Point group	symmetry elements	examples
C_1	E	SiBrClFI
C_2	E, C_2	H_2O_2
C_s	E, σ	NHF_2
C_{2v}	E, C_2, $2\sigma_v$	H_2O, SO_2Cl_2
C_{3v}	E, C_3, $3\sigma_v$	NH_3, PCl_3,$POCl_3$
$C_{\mu v}$	E, C_μ	CO, HCl, OCS
D_{2h}	E, $3C_2$, $2\sigma_v$, σ_h, i	N_2O_4, B_2H_6
D_{3h}	E, C_3, $3C_2$, $3\sigma_v$, σ_h	BF_3, PCl_5
D_{4h}	E, C_4, $4C_2$, $4\sigma_v$, σ_h, i, S_4, ...	XeF_4, trans-MA_4B_2
$D_{\infty h}$	E, C_∞, $\infty\sigma_v$, ...	H_2, CO_2, C_2H_2
T_d	E, $3C_2$, $4C_3$, 6σ, $3S_4$	CH_4, $SiCl_4$
Oh	E, $6C_2$, $4C_3$, $4S_6$, $3S_4$, i, ...	SF_6

CLASSIFICATION OF POINT GROUPS

1. Groups of low symmetry
   a. C_1 no symmetry at all CHFClBr
   b. C_s only one mirror plane $H_2C{=}CClBr$
   c. C_i only an inversion center HClBrC CBrClH
2. C_n groups (n=2, 3, 4, ...) Contain only one C_n rotational axis

   Example: H_2O_2
3. C_{nv} groups (n=2, 3, 4, ...) Contain C_n and $n\sigma_v$ (a set of n vertical mirror planes: these mirror planes contain the C_n axis)

   Examples:

   H_2O NH_3

   SO_2Cl_2 $SClF_5$

   $ONNi(h^5\text{-}C_5H_5)$

4. C_{nh} groups (n=2, 3, ...)

   $C_n + s_h$ s_h: mirror plane, perpendicular to C_n

   Examples : *trans*-HClC=CHCl

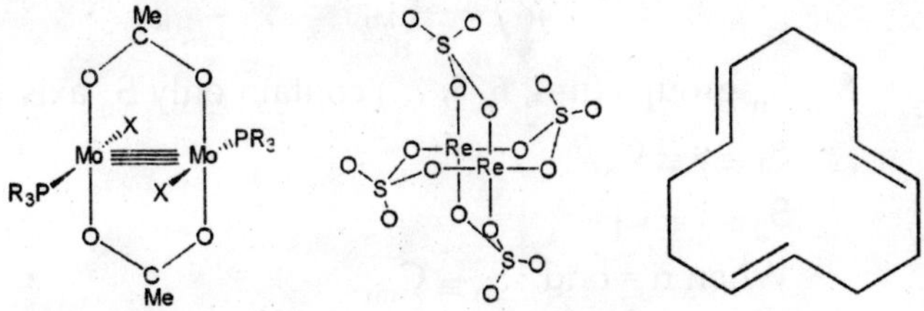

5. D_n groups (n=2, 3, ...)

   C_n + n C_2 A C_n principal rotational axis is accompanied by a set of n C_2 axes perpendicular to it

   example: $Co(H_2NCH_2CH_2NH_2)_3$

6. D_{nh} groups (n=2, 3, 4, ...)

   $D_n + \sigma_h$ σ_h: horizontal mirror plane, perpendicular to the principal axis C_n

   → generating nσ_v mirror planes

   Examples:

Trigonal prism	pentagonal prism,
Trigonal-bipyramidal	pentagonal-bipyramidal

   $PtCl_4^{2-}$

   C_6H_6

   BF_3

   PCl_5

7. D_{nd} groups (n=2, 3, 4, ...) D_n + a set of n vertical mirror planes (σ_v) *do not have* s_h *mirror plane*

   Examples : $R_3W{\equiv}WR_3$

   Ferrocene

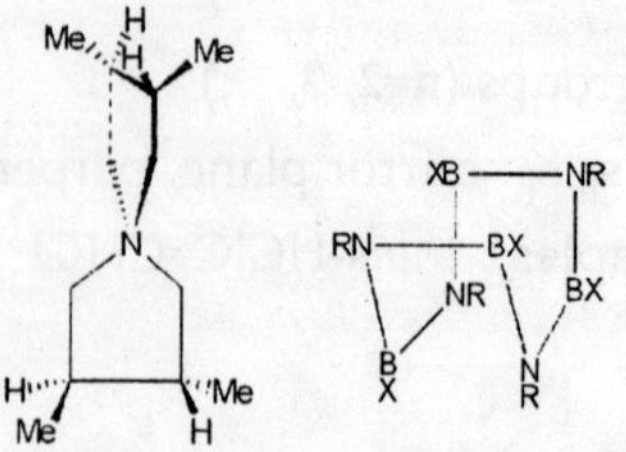

8. S_n groups (n=4, 6, 8, ...) contain only S_n axis

   $S_1 \equiv s \equiv C_s$

   $S_2 \equiv i \equiv C_I$

   When n = odd, $S_n \equiv C_{nh}$

   n can only be even

9. High symmetry point groups Tetrahedral point groups

   T point group : $4C_3 + 3C_2$

   T_d point group : $T + 6\sigma_d$ (CH_4, $SiCl_4$, etc...)

   T_h point group : T + i

   a. Octahedral point groups

      O point group: $4C_3 + 6C_2$

      Oh point group: O + i

   b. Icosahedral point groups

      I point group : $6C_5$

      I_h point group : I + i

10. Point groups for linear molecules

    A-B

    A-B-C

    A-A

    A-B-A

    CO, CO_2, N_2

Some typical shapes of point groups: Procedure in identifying point groups.

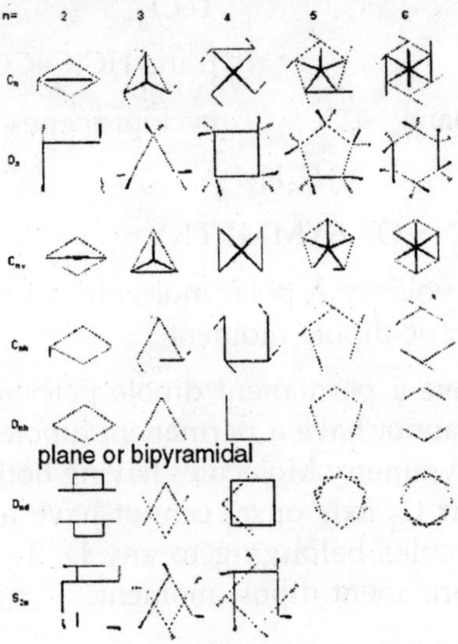

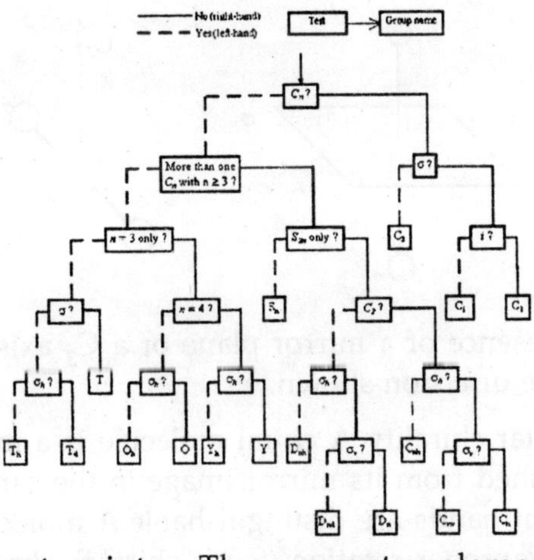

Important points : The symmetry elements used to identify a point group do not form a complete set of symmetry elements for the molecule under consideration. Using the highest symmetry point to assign the molecular symmetry

Exercises :

Mer-$[Fe(CN)_3Cl_3]^{3-}$ $Mo(CN)_8{}^{4-}$

IF_7	P_4
XeO_2F_2	$TeCl_4$
ClSbO	trans-HClC=CClH
Cyclopropane	cyclopropene
C_8H_8	$C_8H_4F_4$

APPLICATIONS OF SYMMETRY

Molecular polarity A polar molecule is a molecule with a permanent electric dipole moment

Cannot have a permanent dipole perpendicular to any mirror plane.Cannot have a permanent dipole perpendicular to any axis of symmetry.Molecules having both a C_n axis and a perpendicular C_2 axis or σ_h cannot have a dipole in any direction.Molecules belonging to any D, T, O or I groups cannot have permanent dipole moment.

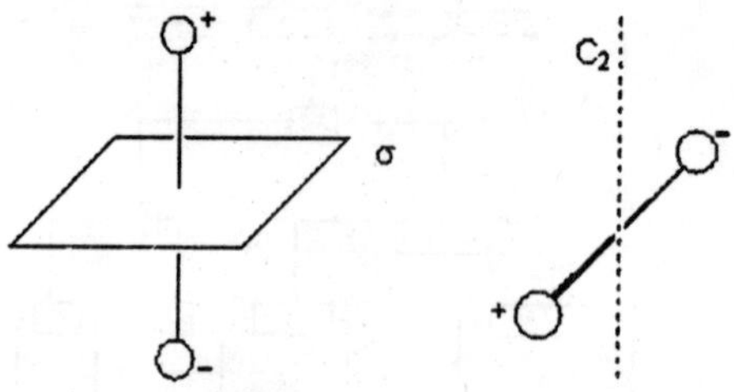

The presence of a mirror plane or a C_2 axis rules out a dipole in the direction shown.

Molecular chirality A chiral molecule is a molecule that is distinguished from its mirror image in the same way that left and right hands are distinguishable.A molecule that has no axis of improper rotation (S_n) is chiral,S_n: including S_1=σ and S_2=i

Chapter 4

Atomic Structure

People have studied reactions and the relationships between substances for centuries but it is only since about 1800 that modern chemistry has developed. Back in 1800, one of the most powerful research tools available was the analytical balance and chemists developed various rules based on mass analysis.

The law of conservation of mass, law of definite proportions and law of multiple proportion were all born of work done at that time. It was John Dalton who was able to link these laws by introducing the atomic theory of matter in which he defined the atom as the simplest unit of matter capable of a separate existence, elements as substances with only one type of atom, compounds containing more than one type of atom and molecules containing two or more atoms. In this respect, Dalton could be described ass the father of modern chemistry.The atom has, however, a longer history. Around 400 B.C., the Greek, Democritus, used the term *atomos* to describe the smallest particle of matter possible. Unfortunately for Democritus, this theory could not be tested at the time and so the concept lay dormant for a couple of thousand years. Bad luck Democritus.In the 1800's, chemists made great strides forward and during this time atomic masses, molecular masses were assigned and new laws, for example, the kinetic theory of gasses were proposed.

	relative mass	relative charge
proton	1	+1
neutron	1	0
electron	1/1836	-1

THE NUCLEUS

The nucleus is at the centre of the atom and contains the protons and neutrons. Protons and neutrons are collectively known as *nucleons.* Virtually all the mass of the atom is concentrated in the nucleus, because the electrons weigh so little.

Working out the numbers of protons and neutrons

No of protons = ATOMIC NUMBER of the atom

The atomic number is also given the more descriptive name of *proton number.*

No of protons + no of neutrons = MASS NUMBER of the atom

The mass number is also called the *nucleon number.* This information can be given simply in the form:

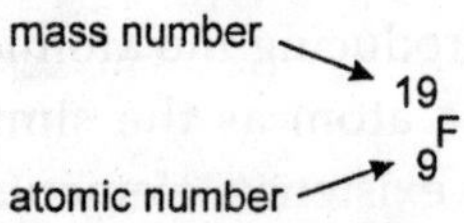

The atomic number counts the number of protons (9); the mass number counts protons + neutrons (19). If there are 9 protons, there must be 10 neutrons for the total to add up to 19. The atomic number is tied to the position of the element in the Periodic Table and therefore the number of protons defines what sort of element you are talking about. So if an atom has 8 protons (atomic number = 8), it must be oxygen. If an atom has 12 protons (atomic number = 12), it must be magnesium. Similarly, every chlorine atom (atomic number = 17) has 17 protons; every uranium atom (atomic number = 92) has 92 protons.

Quantum numbers describe the characteristics of the electron wave and by defining values for these we can indicate the nature of that electron wave including its energy and to a degree its spatial occupation. If this all seems rather complicated, don't worry, it is, but I hope it will become clearer soon.

If you imagine an atom, we wish to indicate where the electron is within that atom. We are also dealing with a three-dimensional (maybe four-dimensional) system. We certainly need to have three values to express the three-dimensional nature of our system and for that we need three quantum numbers. Perhaps the first thing we need to do is indicate, on average, how far away from the nucleus the electron is. Higher energy electrons will be further from the nucleus than lower energy ones. And we can express the energy of an electron using n, the principal quantum number. The greater the value of n, the further away (on average), the electron will be from the nucleus.

PRINCIPAL QUANTUM NUMBER n

This is the easy one. Allowable values of n are positive whole numbers, 1, 2, 3, 4, ... Remember that we are still dealing with probability here but, on average, an electron with a large value of n will be further away from the nucleus than an electron with a small value.The second quantum number, called the subsidiary or angular momentum quantum number is given the symbol l and basically gives us the shape of the orbital. Allowable values of l are from zero up to (n-1). So for n = 1, l= 0 but for n = 2, l= 0, 1 and so on. Each allowable value of l defines the shape of a set of orbitals. When n = 1, there is only one value (0) for l and hence only one orbital with a certain (spherical) shape. For n = 2, however we have two possible values of l (0 and 1) and so we have two possible shapes and two types of orbitals, when l = 0, the orbital is an "s" orbital and when l = 1 we have "p" orbitals.The s orbitals will be spherical whilst the p orbitals, well best look at a diagram, but they have two lobes centred about the nucleus. Together, the values of n and l define a subshell. So for n = 2, we have a subshell consisting of two types of orbital, s and p types. This will be given the label 2s and 2p respectively.Now we get the third quantum number, m_l, the magnetic quantum number. Allowable values of the magnetic quantum number are -l to +l and this number gives us information about the orientation in space of the orbital. If l = 1 then values of m_l are -1, 0, +1. Similarly, if l = 2 the m_l = -2, -1, 0, +1, +2. Now this is interesting. We have partially defined an orbital using n and l. But when l = 1, we have three possible values for ml and this tells us that there are three orbitals with n= 2, l = 1. We have already labelled these orbital as p orbitals and so given that we have three values for ml for this (set of) orbital(s), there must be three p orbitals each orientated differently in space.Of course, when l = 2, we have five possible values of ml, -2, -1, 0, +1, +2. So we have five orbitals each orientated differently n space and these are d orbitals. *How many orbitals will we have when l = 3 ?* (These will be the f orbitals).Now we have defined orbitals in terms of their distance form the nucleus, their shape and their orientation in space and we have seen that for allowable values of n, we can have s, p and d (and f and g) orbitals.

What do these look like ? That's a very interesting question and one to which there are several answers. We'll take the simple way out. It's important to remember that we are taking about probability. So any answer we give to this question must involve probability. What's the probability of finding the electron in this given space ? We must set a defined probability limit, lets say 90%. We can now say that there is a 90% probability that the electron will be within this defined shape. This might be seen as 90% of the time or 90% of the electron, these are terms are interchangeable.

So given this proviso, the shapes of the s, p and d orbitals are discussed later.When we come to the electron we need an extra number. It has been observed that electrons behave as small magnets and it has been suggested that electrons possess the property of spin. We can accommodate this by including a fourth quantum number, m_s, the spin quantum number. The allowable values of this quantum number are easy to remember, it can either by +1/2 or -1/2. Where two electron have the same value for the spin quantum number they are said to have parallel spin and when they are different they have opposed spin.

And now we come to Pauli's Exclusion Principle.

Put bluntly, this states that "No two electrons in one atom can have the same values for all four quantum numbers". This essentially means that a maximum of only two electrons can occupy a single orbital. When two electrons occupy an orbital they must have opposed spin (i.e. different values for the spin quantum number).

ELECTRONIC CONFIGURATION

Each orbital can contain a maximum of two electrons each. So an s orbital can contain two electron as can each p orbital. Hence the set of three p orbitals can contain a maximum of six electrons. We can express this mathematically, "The maximum number of electrons which can occupy a principal quantum group (or shell) is $2n^2$". For $n = 1$, we get 2 electrons, when $n = 2$ we get eight electrons and when $n = 3$ we get 18 electrons and so on. Electrons will enter orbitals at the lowest energy possible. Thus a 1s orbital will be filled before a 2s.

The order of filling orbitals is often shown in text books by way of a diagram, however, a list will do just as well; 1s, 2s, 2p, 3s, 3p, 4s, 3d, 4p, 5s...that will be sufficient for our needs at this level. Notice that the 3d orbitals are filled before the 4p.In addition to filling in terms of energies of orbitals, electrons will also enter orbitals singularly. In other words, if we take the 2p orbitals for example, each will be occupied by a single electron before electrons begin to pair up. Why do you think this is so ? We can explain this by considering the charge on the electron. By placing two like charges together in a confined region of space we set up some repulsion (inter-electron repulsion). The two electrons will effectively repel one another and so their energies will be increased slightly. The aim is always to get to lowest energy and so the electrons will not pair up until forced to do so.

Hund's Rule

As electrons enter a given subgroup, the order of occupation of energy levels (orbitals) is given by Hund's Rule.*Within a given subgroup, each further electron enters a new orbital until all the orbitals are each occupied by one electron only. The electrons in the singly occupied orbitals have parallel spin. Further electrons then enter the same orbitals to complete the pair, each having anti-parallel spin.*

Aufbau Principle

Electrons enter at lowest energy possible.If you take a look at the electronic configuration of elements Z = 24 (Cr) and Z= 29 (Cu).

- Cr = $1s^2, 2s^2, 2p^6, 3s^2, 3p^6, 4s^1, 3d^5$
- Cu = $1s^2, 2s^2, 2p^6, 3s^2, 3p^6, 4s^1, 3d^{10}$

In both these cases, the 4s orbital is only partially full and yet the 3d orbitals have also been filled (at least partially). We can explain this anomaly in terms of the apparent enhanced stability of either half-filled or fully-filled sets of orbitals. It has been shown that atoms with half- or fully-filled sets of orbitals gain additional stability and so with Cr and Cu, we witness a rearrangement of electrons to produce this effect.The electronic configuration of the elements is extremely important as it is this which, by and large, dictates the chemistry of the element.

ISOTOPES

The number of neutrons in an atom can vary within small limits. For example, there are three kinds of carbon atom ^{12}C, ^{13}C and ^{14}C. They all have the same number of protons, but the number of neutrons varies.

	protons	neutrons	mass number
carbon-12	6	6	12
carbon-13	6	7	13
carbon-14	6	8	14

These different atoms of carbon are called *isotopes*. The fact that they have varying numbers of neutrons makes no difference whatsoever to the chemical reactions of the carbon.Isotopes are atoms which have the same atomic number but different mass numbers. They have the same number of protons but different numbers of neutrons.

THE ELECTRONS

Atoms are electrically neutral, and the positiveness of the protons is balanced by the negativeness of the electrons. It follows that in a neutral atom:

no of electrons = no of protons

So, if an oxygen atom (atomic number = 8) has 8 protons, it must also have 8 electrons; if a chlorine atom (atomic number = 17) has 17 protons, it must also have 17 electrons.

The Arrangement of the Electrons

The electrons are found at considerable distances from the nucleus in a series of levels called energy levels. Each energy level can only hold a certain number of electrons. The first level (nearest the nucleus) will only hold 2 electrons, the second holds 8, and the third also seems to be full when it has 8 electrons. At GCSE you stop there because the pattern gets more complicated after that.

These levels can be thought of as getting progressively further from the nucleus. Electrons will always go into the lowest possible energy level (nearest the nucleus) - provided there is space.

- Look up the atomic number in the Periodic Table - making sure that you choose the right number if two numbers are given. The atomic number will always be the smaller one.
- This tells you the number of protons, and hence the number of electrons.
- Arrange the electrons in levels, always filling up an inner level before you go to an outer one.

THE ELECTRONIC ARRANGEMENTS OF THE FIRST 20 ELEMENTS

H 1							He 2
Li 2,1	Be 2,2	B 2,3	C 2,4	N 2,5	O 2,6	F 2,7	Ne 2,8
Na 2,8,1	Mg 2,8,2	Al 2,8,3	Si 2,8,4	P 2,8,5	S 2,8,6	Cl 2,8,7	Ar 2,8,8
K 2,8,8,1	Ca 2,8,8,2						

After this the pattern alters as you enter the transition series in the Periodic Table.

Two Important Generalizations

The number of electrons in the outer level is the same as the group number. (Except with helium which has only 2 electrons. The noble gases are also usually called group 0 - not group 8.) This pattern extends throughout the Periodic Table for the main groups (i.e. not including the transition elements).So if you know that barium is in group 2, it has 2 electrons in its outer level; iodine (group 7) has 7 electrons in its outer level; lead (group 4) has 4 electrons in its outer level. Noble gases have full outer levels. This generalisation will need modifying for A'level purposes.

Dots-and-Crosses Diagrams

In any introductory chemistry course you will have come across the electronic structures of hydrogen and carbon, for example, drawn as:

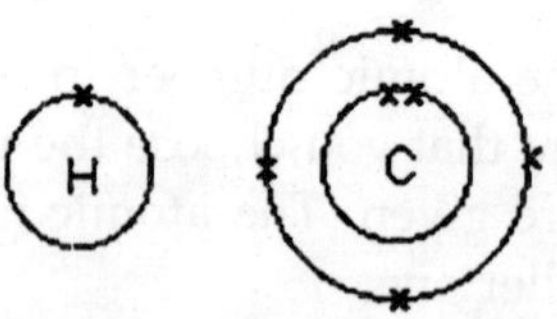

The circles show energy levels - representing increasing distances from the nucleus. You could straighten the circles out and draw the electronic structure as a simple energy diagram.Carbon, for example, would look like this:

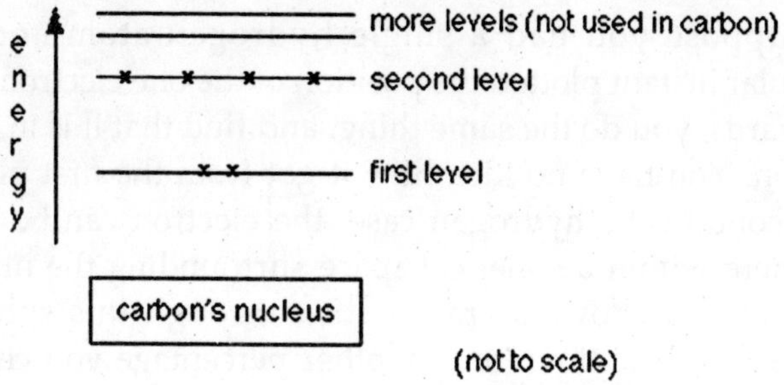

Thinking of the arrangement of the electrons in this way makes a useful bridge to the A'level view.

ATOMIC ORBITALS

Atomic orbitals are in a way that makes them understandable for introductory courses and its equivalents. It explores s and p orbitals in some detail, including their shapes and energies. d orbitals are described only in terms of their energy, and f orbitals only get a passing mention.

ORBITALS AND ORBITS

When the a planet moves around the sun, you can plot a definite path for it which is called an orbit. A simple view of the atom looks similar and you may have pictured the electrons as orbiting around the nucleus. The truth is different, and electrons in fact inhabit regions of space known as *orbitals.*Orbits and orbitals sound similar, but they have quite different meanings. It is essential that you understand the difference between them.

The *Heisenberg Uncertainty Principle* says - loosely - that you can't know with certainty both where an electron is and where it's going next. It is impossible to define with absolute precision, at the same time, both the position and the momentum of an electron.

That makes it impossible to plot an orbit for an electron around a nucleus. Is this a big problem? No. If something is impossible, you have to accept it and find a way around it.

Hydrogen's Electron - the 1s Orbital

Suppose you had a single hydrogen atom and at a particular instant plotted the position of the one electron. Soon afterwards, you do the same thing, and find that it is in a new position. You have no idea how it got from the first place to the second.In the hydrogen case, the electron can be found anywhere within a spherical space surrounding the nucleus. The diagram shows a *cross-section* through this spherical space.95% of the time (or any other percentage you choose), the electron will be found within a fairly easily defined region of space quite close to the nucleus. Such a region of space is called an *orbital*. You can think of an orbital as being the region of space in which the electron lives.What is the electron doing in the orbital? We don't know, we can't know, and so we just ignore the problem! All you can say is that if an electron is in a particular orbital it will have a particular definable energy.Each orbital has a name.The orbital occupied by the hydrogen electron is called a *1s orbital.*

The *"1"* represents the fact that the orbital is in the energy level closest to the nucleus. The *"s"* tells you about the shape of the orbital. s orbitals are spherically symmetric around the nucleus - in each case, like a hollow ball made of rather chunky material with the nucleus at its centre.

The orbital on the left is a *2s orbital.* This is similar to a 1s orbital except that the region where there is the greatest chance of finding the electron is further from the nucleus - this is an orbital at the second energy level.If you look carefully, you will notice that there is another region of slightly higher electron density (where the dots are thicker) nearer the nucleus. ("Electron density" is another way of talking about how likely you are to find an electron at a particular place.)2s (and 3s, 4s, etc) electrons spend some of their time closer to the nucleus than you might expect. The effect of this is to slightly reduce the energy of electrons in s orbitals. The nearer the nucleus the electrons get, the lower their energy.3s, 4s (etc) orbitals get progressively further from the nucleus.

p Orbitals

Not all electrons inhabit s orbitals (in fact, very few electrons live in s orbitals). At the first energy level, the only orbital available to electrons is the 1s orbital, but at the second level, as well as a 2s orbital, there are also orbitals called *2p orbitals.*A p orbital is rather like 2 identical balloons tied together at the nucleus. The diagram on the right is a cross-section through that 3-dimensional region of space. Once again, the orbital shows where there is a 95% chance of finding a particular electron.Unlike an s orbital, a p orbital points in a particular direction.At any one energy level it is possible to have three absolutely equivalent p orbitals pointing mutually at right angles to each other. These are arbitrarily given the symbols p_x, p_y and p_z. This is simply for convenience - what you might think of as the x, y or z direction changes constantly as the atom tumbles in space.The p orbitals at the second energy level are called $2p_x$, $2p_y$ and $2p_z$. There are similar orbitals at subsequent levels - $3p_x$, $3p_y$, $3p_z$, $4p_x$, $4p_y$, $4p_z$ and so on.All levels except for the first level have p orbitals. At the higher levels the lobes get more elongated, with the most likely place to find the electron more distant from the nucleus.

d and f Orbitals

In addition to s and p orbitals, there are two other sets of orbitals which become available for electrons to inhabit at higher energy levels. At the third level, there is a set of five d orbitals (with complicated shapes and names) as well as the 3s and 3p orbitals ($3p_x$, $3p_y$, $3p_z$). At the third level there are a total of nine orbitals altogether.At the fourth level, as well the 4s and 4p and 4d orbitals there are an additional seven f orbitals - 16 orbitals in all. s, p, d and f orbitals are then available at all higher energy levels as well.For the moment, you need to be aware that there are sets of five d orbitals at levels from the third level upwards, but you probably won't be expected to draw them or name them. Apart from a passing reference, you won't come across f orbitals at all.

Fitting Electrons into Orbitals

You can think of an atom as a very bizarre house (like an inverted pyramid!) - with the nucleus living on the ground floor, and then various rooms (orbitals) on the higher floors occupied by the electrons. On the first floor there is only 1 room (the 1s orbital); on the second floor there are 4 rooms (the 2s, $2p_x$, $2p_y$ and $2p_z$ orbitals); on the third floor there are 9 rooms (one 3s orbital, three 3p orbitals and five 3d orbitals); and so on. But the rooms aren't very big . . . Each orbital can only hold 2 electrons.

A convenient way of showing the orbitals that the electrons live in is to draw "electrons-in-boxes".

"Electrons-in-Boxes"

Orbitals can be represented as boxes with the electrons in them shown as arrows. Often an up-arrow and a down-arrow are used to show that the electrons are in some way different.

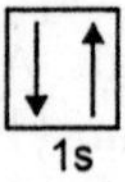

A 1s orbital holding 2 electrons would be drawn as shown on the right, but it can be written even more quickly as $1s^2$. This is read as "one s two" - not as "one s squared".You mustn't confuse the two numbers in this notation:

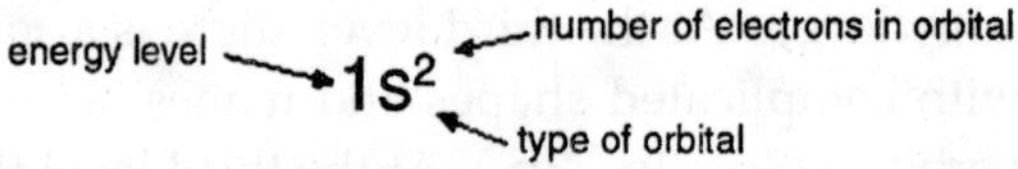

The Order of Filling Orbitals

Electrons fill low energy orbitals (closer to the nucleus) before they fill higher energy ones. Where there is a choice between orbitals of equal energy, they fill the orbitals singly as far as possible.This filling of orbitals singly where possible is known as *Hund's rule*. It only applies where the orbitals have exactly the same energies (as with p orbitals, for example), and helps to minimise the repulsions between electrons and so makes the atom more stable.

The diagram (not to scale) summarises the energies of the orbitals up to the 4p level.

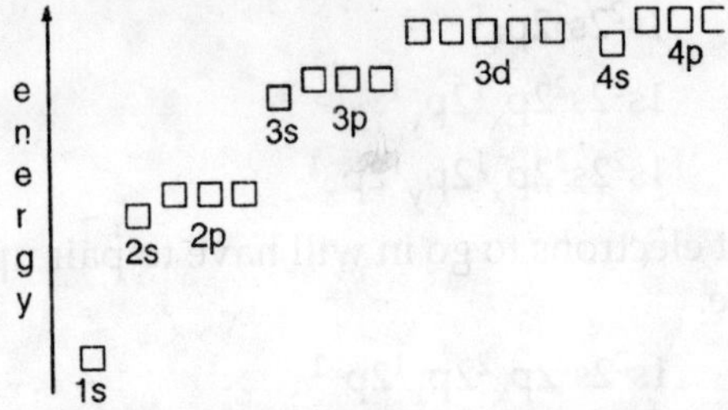

Notice that the s orbital always has a slightly lower energy than the p orbitals at the same energy level, so the s orbital always fills with electrons before the corresponding p orbitals.The real oddity is the position of the 3d orbitals. They are at a slightly higher level than the 4s - and so it is the 4s orbital which will fill first, followed by all the 3d orbitals and then the 4p orbitals. Similar confusion occurs at higher levels, with so much overlap between the energy levels that the 4f orbitals don't fill until after the 6s, for example.

ELECTRONIC STRUCTURES

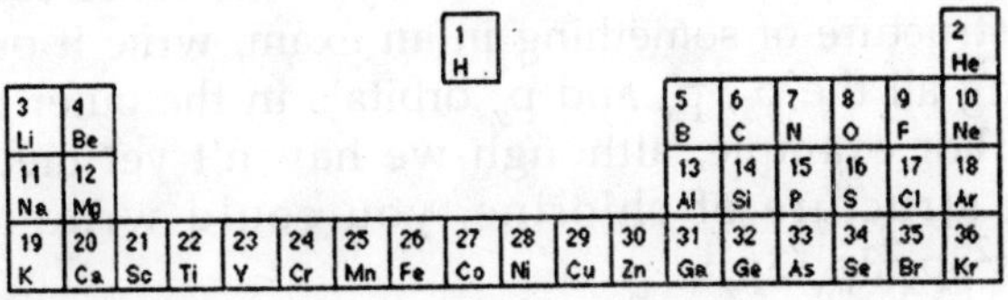

The First Period

Hydrogen has its only electron in the 1s orbital - $1s^1$, and at helium the first level is completely full - $1s^2$.

The Second Period

Now we need to start filling the second level, and hence start the second period. Lithium's electron goes into the 2s orbital because that has a lower energy than the 2p orbitals. Lithium has an electronic structure of $1s^2 2s^1$. Beryllium adds a second electron to this same level - $1s^2 2s^2$.

Now the 2p levels start to fill. These levels all have the same energy, and so the electrons go in singly at first.

B $1s^2 2s^2 2p_x^{\ 1}$

C $1s^2 2s^2 2p_x^{\ 1} 2p_y^{\ 1}$

N $1s^2 2s^2 2p_x^{\ 1} 2p_y^{\ 1} 2p_z^{\ 1}$

The next electrons to go in will have to pair up with those already there.

O $1s^2 2s^2 2p_x^{\ 2} 2p_y^{\ 1} 2p_z^{\ 1}$

F $1s^2 2s^2 2p_x^{\ 2} 2p_y^{\ 2} 2p_z^{\ 1}$

Ne $1s^2 2s^2 2p_x^{\ 2} 2p_y^{\ 2} 2p_z^{\ 2}$

The full electronic structures of atoms as the number of electrons increases. There are two ways around this,

All the various p electrons can be lumped together. For example, fluorine could be written as $1s^2 2s^2 2p^5$, and neon as $1s^2 2s^2 2p^6$.This is what is normally done if the electrons are in an inner layer. If the electrons are in the bonding level (those on the outside of the atom), they are sometimes written in shorthand, sometimes in full. Don't worry about this. Be prepared to meet either version, but if you are asked for the electronic structure of something in an exam, write it out in full showing all the p_x, p_y and p_z orbitals in the outer level separately.For example, although we haven't yet met the electronic structure of chlorine, you could write it as $1s^2 2s^2 2p^6 3s^2 3p_x^{\ 2} 3p_y^{\ 2} 3p_z^{\ 1}$.

The 2p electrons are all lumped together whereas the 3p ones are shown in full. The logic is that the 3p electrons will be involved in bonding because they are on the outside of the atom, whereas the 2p electrons are buried deep in the atom and aren't really of any interest.

All the inner electrons together using, for example, the symbol [Ne]. In this context, [Ne] means *the electronic structure of neon* - in other words: $1s^2 2s^2 2p_x^{\ 2} 2p_y^{\ 2} 2p_z^{\ 2}$ You wouldn't do this with helium because it takes longer to write [He] than it does $1s^2$.On this basis the structure of chlorine would be written $[Ne]3s^2 3p_x^{\ 2} 3p_y^{\ 2} 3p_z^{\ 1}$.

The Third Period

At neon, all the second level orbitals are full, and so after this we have to start the third period with sodium. The pattern of filling is now exactly the same as in the previous period, except that everything is now happening at the 3-level.

For example:

		short version
Mg	$1s^22s^22p^63s^2$	$[Ne]3s^2$
S	$1s^22s^22p^63s^23p_x{}^23p_y{}^13p_z{}^1$	$[Ne]3s^23p_x{}^23p_y{}^13p_z{}^1$
Ar	$1s^22s^22p^63s^23p_x{}^23p_y{}^23p_z{}^2$	$[Ne]3s^23p_x{}^23p_y{}^23p_z{}^2$

The Beginning of the Fourth Period

At this point the 3-level orbitals aren't all full - the 3d levels haven't been used yet. But if you refer back to the energies of the orbitals, you will see that the next lowest energy orbital is the 4s - so that fills next.

K	$1s^22s^22p^63s^23p^64s^1$
Ca	$1s^22s^22p^63s^23p^64s^2$

There is strong evidence for this in the similarities in the chemistry of elements like sodium ($1s^22s^22p^63s^1$) and potassium ($1s^22s^22p^63s^23p^64s^1$)

The outer electron governs their properties and that electron is in the same sort of orbital in both of the elements. That wouldn't be true if the outer electron in potassium was $3d^1$.

s- and p-block Elements

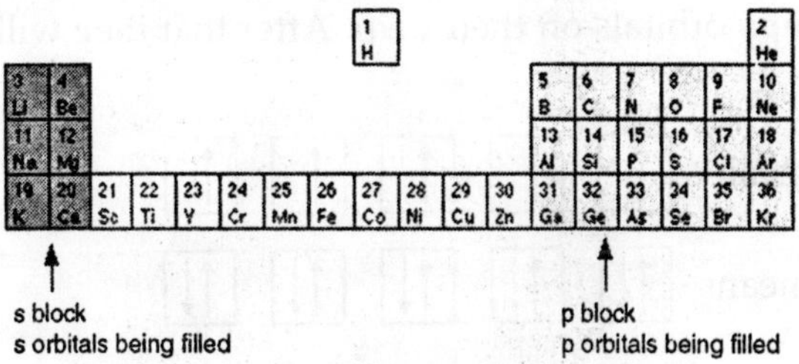

The elements in group 1 of the Periodic Table all have an outer electronic structure of ns^1 (where n is a number between 2 and 7). All group 2 elements have an outer electronic structure of ns^2. Elements in groups 1 and 2 are described as s-block elements.Elements from group 3 across to the noble gases all have their outer electrons in p orbitals. These are then described as p-block elements.

d-block Elements

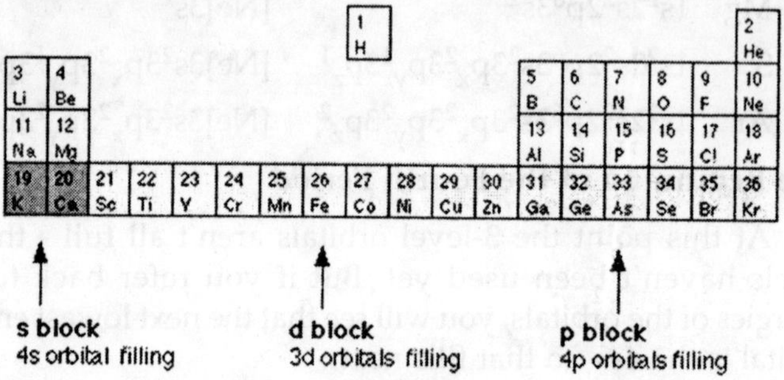

Remember that the 4s orbital has a lower energy than the 3d orbitals and so fills first. Once the 3d orbitals have filled up, the next electrons go into the 4p orbitals as you would expect.d-block elements are elements in which the last electron to be added to the atom is in a d orbital. The first series of these contains the elements from scandium to zinc, which at GCSE you probably called transition elements or transition metals. The terms "transition element" and "d-block element" don't quite have the same meaning, but it doesn't matter in the present context.d electrons are almost always described as, for example, d^5 or d^8 - and not written as separate orbitals. Remember that there are five d orbitals, and that the electrons will inhabit them singly as far as possible. Up to 5 electrons will occupy orbitals on their own. After that they will have to pair up.

d^5 means 

d^8 means 

Notice in what follows that all the 3-level orbitals are written together, even though the 3d electrons are added to the atom after the 4s.

Sc $1s^22s^22p^63s^23p^63d^14s^2$

Ti $1s^22s^22p^63s^23p^63d^24s^2$

V $1s^22s^22p^63s^23p^63d^34s^2$

Cr $1s^22s^22p^63s^23p^63d^54s^1$

Whoops! Chromium breaks the sequence. In chromium, the electrons in the 3d and 4s orbitals rearrange so that there is one electron in each orbital. It would be convenient if the sequence was tidy - but it's not!

Mn $1s^22s^22p^63s^23p^63d^54s^2$ (back to being tidy again)

Fe $1s^22s^22p^63s^23p^63d^64s^2$

Co $1s^22s^22p^63s^23p^63d^74s^2$

Ni $1s^22s^22p^63s^23p^63d^84s^2$

Cu $1s^22s^22p^63s^23p^63d^{10}4s^1$ (another awkward one!)

Zn $1s^22s^22p^63s^23p^63d^{10}4s^2$

And at zinc the process of filling the d orbitals is complete.

FILLING THE REST OF PERIOD 4

The next orbitals to be used are the 4p, and these fill in exactly the same way as the 2p or 3p. We are back now with the p-block elements from gallium to krypton. Bromine, for example, is $1s^22s^22p^63s^23p^63d^{10}4s^24p_x^{\,2}4p_y^{\,2}4p_z^{\,1}$.

The electronic structure of an element from hydrogen to krypton

- Use the Periodic Table to find the atomic number, and hence number of electrons.
- Fill up orbitals in the order 1s, 2s, 2p, 3s, 3p, 4s, 3d, 4p - until you run out of electrons. The 3d is the awkward one - remember that specially. Fill p and d orbitals singly as far as possible before pairing electrons up.

- Remember that chromium and copper have electronic structures which break the pattern in the first row of the d-block.

THE ELECTRONIC STRUCTURE OF BIG S- OR P-BLOCK ELEMENTS

First work out the number of outer electrons. This is quite likely all you will be asked to do anyway.The number of outer electrons is the same as the group number. (The noble gases are a bit of a problem here, because they are normally called group 0 rather then group 8. Helium has 2 outer electrons; the rest have 8.) All elements in group 3, for example, have 3 electrons in their outer level. Fit these electrons into s and p orbitals as necessary. Which level orbitals? Count the periods in the Periodic Table (not forgetting the one with H and He in it).

Iodine is in group 7 and so has 7 outer electrons. It is in the fifth period and so its electrons will be in 5s and 5p orbitals. Iodine has the outer structure $5s^2 5p_x^{\,2} 5p_y^{\,2} 5p_z^{\,1}$.

What about the inner electrons if you need to work them out as well? The 1, 2 and 3 levels will all be full, and so will the 4s, 4p and 4d.

That gives the full structure: $1s^2 2s^2 2p^6 3s^2 3p^6 3d^{10} 4s^2 4p^6 4d^{10} 5s^2 5p_x^{\,2} 5p_y^{\,2} 5p_z^{\,1}$.

When you've finished, count all the electrons to make sure that they come to the same as the atomic number.

Barium is in group 2 and so has 2 outer electrons. It is in the sixth period. Barium has the outer structure $6s^2$.

Including all the inner levels: $1s^2 2s^2 2p^6 3s^2 3p^6 3d^{10} 4s^2 4p^6 4d^{10} 5s^2 5p^6 6s^2$.

It would be easy to include $5d^{10}$ as well by mistake, but the d level always fills *after* the next s level - so 5d fills after 6s just as 3d fills after 4s. As long as you counted the number of electrons you could easily spot this mistake because you would have 10 too many.

Electronic Structures of Ions

Ions are atoms (or groups of atoms) which carry an electric charge because they have either gained or lost one or more electrons. If an atom gains electrons it acquires a negative charge. If it loses electrons, it becomes positively charged.

The electronic structure of s- and p-block ions

Write the electronic structure for the neutral atom, and then add (for a negative ion) or subtract electrons (for a positive ion).

To write the electronic structure for Cl^-:

Cl $1s^22s^22p^63s^23p_x{}^23p_y{}^23p_z{}^1$ but Cl^- has one more electron

Cl^- $1s^22s^22p^63s^23p_x{}^23p_y{}^23p_z{}^2$

To write the electronic structure for O^{2-}:

O $1s^22s^22p_x{}^22p_y{}^12p_z{}^1$ but O^{2-} has two more electrons

O^{2-} $1s^22s^22p_x{}^22p_y{}^22p_z{}^2$

To write the electronic structure for Na^+:

Na $1s^22s^22p^63s^1$ but Na^+ has one less electron

Na^+ $1s^22s^22p^6$

The electronic structure for Ca^{2+}:

Ca $1s^22s^22p^63s^23p^64s^2$ but Ca^{2+} has two less electrons

Ca^{2+} $1s^22s^22p^63s^23p^6$

The Electronic Structure of d-block Ions

Here you are faced with one of the most irritating facts in A'level chemistry! You will recall that the first transition series (from scandium to zinc) is the result of the 3d orbitals being filled after the 4s orbital.

However, once the electrons are established in their orbitals, the energy order changes - and in all the chemistry of the transition elements, the 4s orbital behaves as the outermost, highest energy orbital. The reversed order of the 3d and 4s orbitals only applies to building the atom up in the first place.

In all other respects, the 4s electrons are always the electrons you need to think about first.

When d-block elements form ions, the 4s electrons are lost first.

Provided you remember that, working out the structure of a d-block ion is no different from working out the structure of, say, a sodium ion.

The electronic structure for Cr^{3+}:

Cr $1s^2 2s^2 2p^6 3s^2 3p^6 3d^5 4s^1$

Cr^{3+} $1s^2 2s^2 2p^6 3s^2 3p^6 3d^3$

The 4s electron is lost first followed by two of the 3d electrons.

The *electronic structure for Zn^{2+}:*

Zn $1s^2 2s^2 2p^6 3s^2 3p^6 3d^{10} 4s^2$

Zn^{2+} $1s^2 2s^2 2p^6 3s^2 3p^6 3d^{10}$

This time there is no need to use any of the 3d electrons.

The electronic structure for Fe^{3+}:

Fe $1s^2 2s^2 2p^6 3s^2 3p^6 3d^6 4s^2$

Fe^{3+} $1s^2 2s^2 2p^6 3s^2 3p^6 3d^5$

The 4s electrons are lost first followed by one of the 3d electrons.The rule is quite simple. Take the 4s electrons off first, and then as many 3d electrons as necessary to produce the correct positive charge.

IONISATION ENERGY

The first ionisation energy is the energy required to remove the most loosely held electron from one mole of gaseous atoms to produce 1 mole of gaseous ions each with a charge of 1+.

This is more easily seen in symbol terms.

$$X_{(g)} \longrightarrow X^+_{(g)} + e^-$$

It is the energy needed to carry out this change per mole of X.

The state symbols - (g) - are essential. When you are talking about ionisation energies, everything must be present in the gas state.Ionisation energies are measured in kJ mol^{-1} (kilojoules per mole). They vary in size from 381 (which you would consider very low) up to 2370 (which is very high).All elements have a first ionisation energy - even atoms which don't form positive ions in test tubes. The reason that helium (1st I.E. = 2370 kJ mol^{-1}) doesn't normally form a positive ion is because of the huge amount of energy that would be needed to remove one of its electrons.

Table : Patterns of first ionisation energies in the Periodic Table

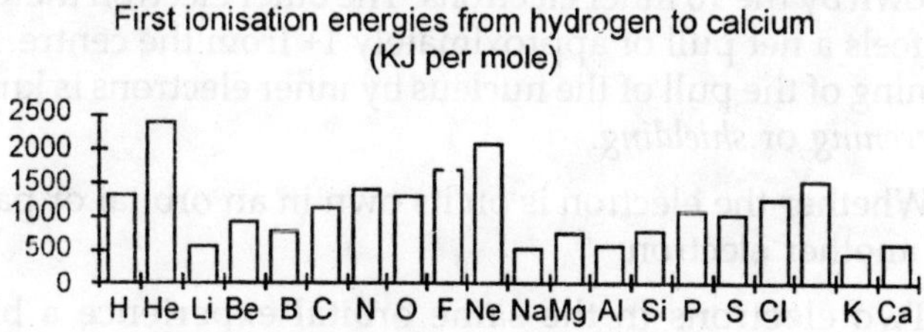

First ionisation energy shows *periodicity.* That means that it varies in a repetitive way as you move through the Periodic Table. For example, look at the pattern from Li to Ne, and then compare it with the identical pattern from Na to Ar.These variations in first ionisation energy can all be explained in terms of the structures of the atoms involved.

Factors Affecting the Size of Ionisation Energy

Ionisation energy is a measure of the energy needed to pull a particular electron away from the attraction of the nucleus. A high value of ionisation energy shows a high attraction between the electron and the nucleus. The size of that attraction will be governed by:

The charge on the nucleus.

The more protons there are in the nucleus, the more positively charged the nucleus is, and the more strongly electrons are attracted to it.

The distance of the electron from the nucleus.

Attraction falls off very rapidly with distance. An electron close to the nucleus will be much more strongly attracted than one further away.

The number of electrons between the outer electrons and the nucleus.

Consider a sodium atom, with the electronic structure 2,8,1. (There's no reason why you can't use this notation if it's useful!)

If the outer electron looks in towards the nucleus, it doesn't see the nucleus sharply. Between it and the nucleus there are the two layers of electrons in the first and second levels. The 11 protons in the sodium's nucleus have their effect cut down by the 10 inner electrons. The outer electron therefore only feels a net pull of approximately 1+ from the centre. This lessening of the pull of the nucleus by inner electrons is known as *screening* or *shielding*.

Whether the electron is on its own in an orbital or paired with another electron.

Two electrons in the same orbital experience a bit of repulsion from each other. This offsets the attraction of the nucleus, so that paired electrons are removed rather more easily than you might expect.

Hydrogen has an electronic structure of $1s^1$. It is a very small atom, and the single electron is close to the nucleus and therefore strongly attracted. There are no electrons screening it from the nucleus and so the ionisation energy is high (1310 kJ mol^{-1}).

Helium has a structure $1s^2$. The electron is being removed from the same orbital as in hydrogen's case. It is close to the nucleus and unscreened. The value of the ionisation energy (2370 kJ mol^{-1}) is much higher than hydrogen, because the nucleus now has 2 protons attracting the electrons instead of 1.

Lithium is $1s^2 2s^1$. Its outer electron is in the second energy level, much more distant from the nucleus. You might argue that that would be offset by the additional proton in the nucleus, but the electron doesn't feel the full pull of the nucleus - it is screened by the $1s^2$ electrons.

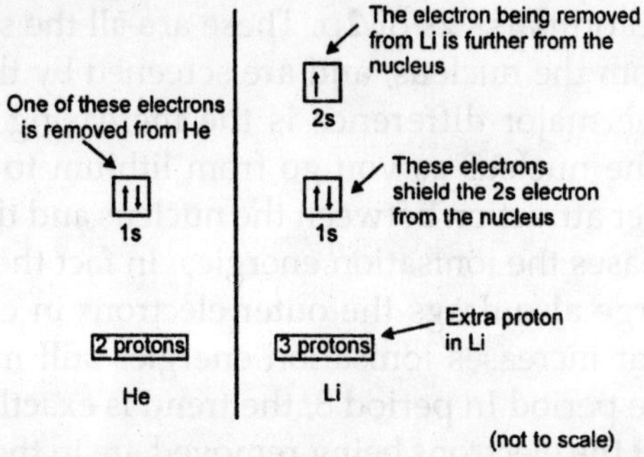

You can think of the electron as feeling a net 1+ pull from the centre (3 protons offset by the two $1s^2$ electrons).

If you compare lithium with hydrogen (instead of with helium), the hydrogen's electron also feels a 1+ pull from the nucleus, but the distance is much greater with lithium. Lithium's first ionisation energy drops to 519 kJ mol^{-1} whereas hydrogen's is 1310 kJ mol^{-1}.

The patterns in periods 2 and 3

Talking through the next 17 atoms one at a time would take ages. We can do it much more neatly by explaining the main trends in these periods, and then accounting for the exceptions to these trends.

The first thing to realise is that the patterns in the two periods are identical - the difference being that the ionisation energies in period 3 are all lower than those in period 2.

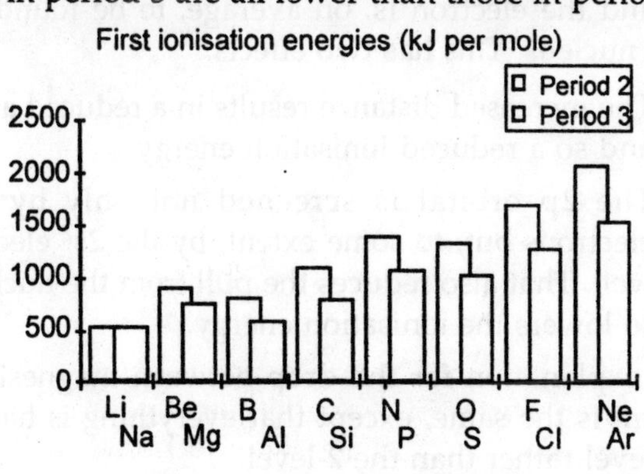

The general trend is for ionisation energies to increase across a period.In the whole of period 2, the outer electrons are in 2-level orbitals - 2s or 2p. These are all the same sort of distances from the nucleus, and are screened by the same $1s^2$ electrons.The major difference is the increasing number of protons in the nucleus as you go from lithium to neon. That causes greater attraction between the nucleus and the electrons and so increases the ionisation energies. In fact the increasing nuclear charge also drags the outer electrons in closer to the nucleus. That increases ionisation energies still more as you go across the period.In period 3, the trend is exactly the same. This time, all the electrons being removed are in the third level and are screened by the $1s^22s^22p^6$ electrons. They all have the same sort of environment, but there is an increasing nuclear charge.

The explanation lies with the structures of boron and aluminium. The outer electron is removed more easily from these atoms than the general trend in their period would suggest.

Be	$1s^22s^2$	1st I.E. = 900 kJ mol^{-1}
B	$1s^22s^22p_x^1$	1st I.E. = 799 kJ mol^{-1}

You might expect the boron value to be more than the beryllium value because of the extra proton. Offsetting that is the fact that boron's outer electron is in a 2p orbital rather than a 2s. 2p orbitals have a slightly higher energy than the 2s orbital, and the electron is, on average, to be found further from the nucleus. This has two effects.

a. The increased distance results in a reduced attraction and so a reduced ionisation energy.

b. The 2p orbital is screened not only by the $1s^2$ electrons but, to some extent, by the $2s^2$ electrons as well. That also reduces the pull from the nucleus and so lowers the ionisation energy.

The explanation for the drop between magnesium and aluminium is the same, except that everything is happening at the 3-level rather than the 2-level.

Mg $1s^2 2s^2 2p^6 3s^2$ 1st I.E. = 736 kJ mol^{-1}

Al $1s^2 2s^2 2p^6 3s^2 3p_x^{\ 1}$ 1st I.E. = 577 kJ mol^{-1}

The 3p electron in aluminium is slightly more distant from the nucleus than the 3s, and partially screened by the $3s^2$ electrons as well as the inner electrons. Both of these factors offset the effect of the extra proton.

Once again, you might expect the ionisation energy of the group 6 element to be higher than that of group 5 because of the extra proton. What is offsetting it this time?

N $1s^2 2s^2 2p_x^{\ 1} 2p_y^{\ 1} 2p_z^{\ 1}$ 1st I.E. = 1400 kJ mol^{-1}

O $1s^2 2s^2 2p_x^{\ 2} 2p_y^{\ 1} 2p_z^{\ 1}$ 1st I.E. = 1310 kJ mol^{-1}

The screening is identical (from the $1s^2$ and, to some extent, from the $2s^2$ electrons), and the electron is being removed from an identical orbital.

The difference is that in the oxygen case the electron being removed is one of the $2p_x^{\ 2}$ pair. The repulsion between the two electrons in the same orbital means that the electron is easier to remove than it would otherwise be.The drop in ionisation energy at sulphur is accounted for in the same way.

As you go down a group in the Periodic Table ionisation energies generally fall. You have already seen evidence of this in the fact that the ionisation energies in period 3 are all less than those in period 2.Taking Group 1 as a typical example:

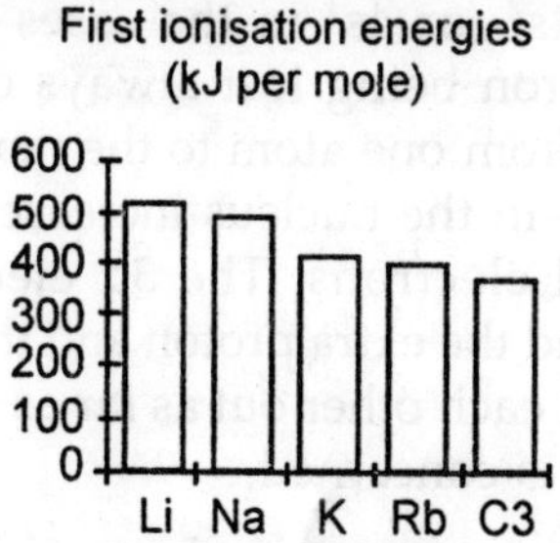

There are 11 protons in a sodium atom but only 3 in a lithium atom, so the nuclear charge is much greater. You might have expected a much larger ionisation energy in sodium, but offsetting the nuclear charge is a greater distance from the nucleus and more screening.

Li	$1s^22s^1$	1st I.E. = 519 kJ mol^{-1}
Na	$1s^22s^22p^63s^1$	1st I.E. = 494 kJ mol^{-1}

Lithium's outer electron is in the second level, and only has the $1s^2$ electrons to screen it. The $2s^1$ electron feels the pull of 3 protons screened by 2 electrons - a net pull from the centre of 1+.

The sodium's outer electron is in the third level, and is screened from the 11 protons in the nucleus by a total of 10 inner electrons. The $3s^1$ electron also feels a net pull of 1+ from the centre of the atom. In other words, the effect of the extra protons is compensated for by the effect of the extra screening electrons. The only factor left is the extra distance between the outer electron and the nucleus in sodium's case. That lowers the ionisation energy.Similar explanations hold as you go down the rest of this group - or, indeed, any other group.

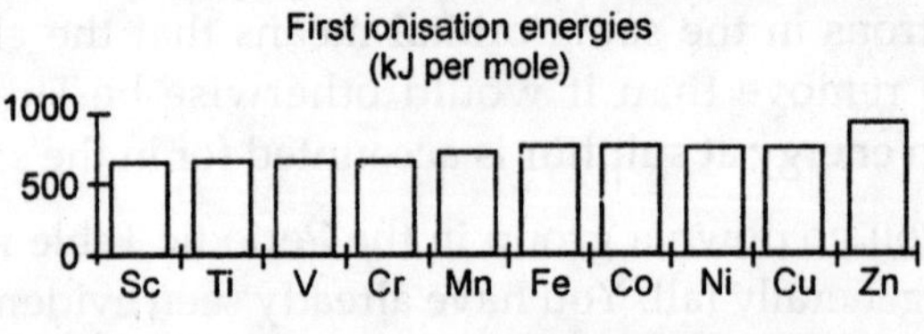

Apart from zinc at the end, the other ionisation energies are all much the same.All of these elements have an electronic structure $[Ar]3d^n4s^2$ (or $4s^1$ in the cases of chromium and copper). The electron being lost always comes from the 4s orbital.As you go from one atom to the next in the series, the number of protons in the nucleus increases, but so also does the number of 3d electrons. The 3d electrons have some screening effect, and the extra proton and the extra 3d electron more or less cancel each other out as far as attraction from the centre of the atom is concerned.

The rise at zinc is easy to explain.

Cu	$[Ar]3d^{10}4s^1$	1st I.E. = 745 kJ mol^{-1}
Zn	$[Ar]3d^{10}4s^2$	1st I.E. = 908 kJ mol^{-1}

In each case, the electron is coming from the same orbital, with identical screening, but the zinc has one extra proton in the nucleus and so the attraction is greater. There will be a degree of repulsion between the paired up electrons in the 4s orbital, but in this case it obviously isn't enough to outweigh the effect of the extra proton.

Ionisation Energies and Reactivity

The lower the ionisation energy, the more easily this change happens:

$$X_{(g)} \longrightarrow X^{+}_{(g)} + e^{-}$$

You can explain the increase in reactivity of the Group 1 metals (Li, Na, K, Rb, Cs) as you go down the group in terms of the fall in ionisation energy. Whatever these metals react with, they have to form positive ions in the process, and so the lower the ionisation energy, the more easily those ions will form.The danger with this approach is that the formation of the positive ion is only one stage in a multi-step process.For example, you wouldn't be starting with gaseous atoms; nor would you end up with gaseous positive ions - you would end up with ions in a solid or in solution. The energy changes in these processes also vary from element to element. Ideally you need to consider the whole picture and not just one small part of it.However, the ionisation energies of the elements are going to be major contributing factors towards the *activation energy* of the reactions. Remember that activation energy is the minimum energy needed before a reaction will take place. The lower the activation energy, the faster the reaction will be - irrespective of what the *overall* energy changes in the reaction are.

The fall in ionisation energy as you go down a group will lead to lower activation energies and therefore faster reactions.

SUCCESSIVE IONISATION ENERGIES

The second, third, (etc) ionisation energy means, and then looks at patterns in successive ionisation energies for selected elements. It assumes that you understand about first ionisation energy.

Second ionisation energy is defined by the equation:

$$X^{+}_{(g)} \longrightarrow X^{2+}_{(g)} + e^{-}$$

It is the energy needed to remove a second electron from each ion in 1 mole of gaseous 1+ ions to give gaseous 2+ ions.

More ionisation energies

You can then have as many successive ionisation energies as there are electrons in the original atom.

The first four ionisation energies of aluminium, for example, are given by

$$Al_{(g)} \longrightarrow Al^{+}_{(g)} + e^{-} \quad \text{1st I.E.} = 577 \text{ kJ mol}^{-1}$$

$$Al^{+}_{(g)} \longrightarrow Al^{2+}_{(g)} + e^{-} \quad \text{2nd I.E.} = 1820 \text{ kJ mol}^{-1}$$

$$Al^{2+}_{(g)} \longrightarrow Al^{3+}_{(g)} + e^{-} \quad \text{3rd I.E.} = 2740 \text{ kJ mol}^{-1}$$

$$Al^{3+}_{(g)} \longrightarrow Al^{4+}_{(g)} + e^{-} \quad \text{4th I.E.} = 11600 \text{ kJ mol}^{-1}$$

In order to form an $Al^{3+}_{(g)}$ ion from $Al_{(g)}$ you would have to supply:

$577 + 1820 + 2740 = 5137 \text{ kJ mol}^{-1}$

That's a lot of energy. Why, then, does aluminium form Al^{3+} ions?

It can only form them if it can get that energy back from somewhere, and whether that's feasible depends on what it is reacting with.

For example, if aluminium reacts with fluorine or oxygen, it can recover that energy in various changes involving the fluorine or oxygen - and so aluminium fluoride or aluminium oxide contain Al^{3+} ions.

If it reacts with chlorine, it can't recover sufficient energy, and so solid anhydrous aluminium chloride isn't actually ionic - instead, it forms covalent bonds.

Why doesn't aluminium form an Al^{4+} ion? The fourth ionisation energy is huge compared with the first three, and there is nothing that aluminium can react with which would enable it to recover that amount of extra energy.Once you have removed the first electron you are left with a positive ion. Trying to remove a negative electron from a positive ion is going to be more difficult than removing it from an atom. Removing an electron from a 2+ or 3+ (etc) ion is going to be progressively more difficult.The electronic structure of aluminium is $1s^2 2s^2 2p^6 3s^2 3p_x^1$. The first three electrons to be removed are the three electrons in the 3p and 3s orbitals. Once they've gone, the fourth electron is removed from the 2p level - much closer to the nucleus, and only screened by the $1s^2$ (and to some extent the $2s^2$) electrons.

This big jump between two successive ionisation energies is typical of suddenly breaking in to an inner level. You can use this to work out which group of the Periodic Table an element is in from its successive ionisation energies.

Magnesium ($1s^2 2s^2 2p^6 3s^2$) is in group 2 of the Periodic Table and has successive ionisation energies:

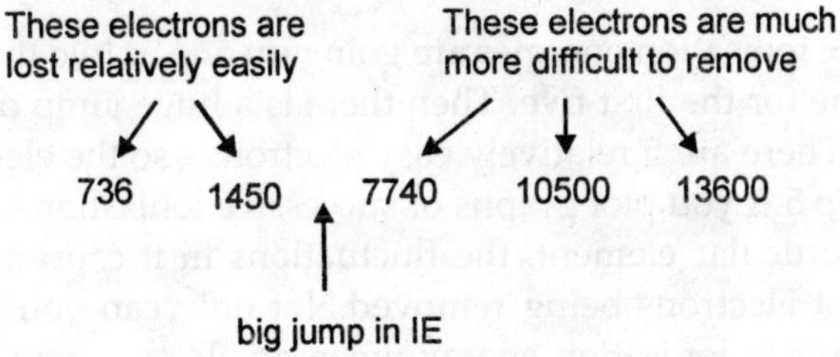

Here the big jump occurs after the second ionisation energy. It means that there are 2 electrons which are relatively easy to remove (the $3s^2$ electrons), while the third one is much more difficult (because it comes from an inner level - closer to the nucleus and with less screening).Silicon ($1s^2 2s^2 2p^6 3s^2 3p_x^1 3p_y^1$) is in group 4 of the Periodic Table and has successive ionisation energies:

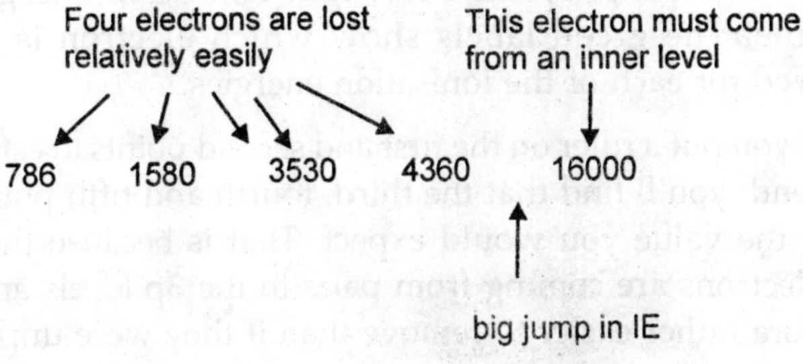

Here the big jump comes after the fourth electron has been removed. The first 4 electrons are coming from the 3-level orbitals; the fifth from the 2-level.

The lesson from all this:

Count the easy electrons - those up to (but not including) the big jump. That is the same as the group number.

Another example:

Decide which group an atom is in if it has successive ionisation energies:

1060 1900 2920 4960 6280 21200

↑ big jump in IE after the fifth electron

The ionisation energies are going up one or two thousand at a time for the first five. Then there is a huge jump of about 15000. There are 5 relatively easy electrons - so the element is in group 5.If you plot graphs of successive ionisation energies for a particular element, the fluctuations in it caused by the different electrons being removed.Not only can you see the big jumps in ionisation energy when an electron comes from an inner level, but you can also see the minor fluctuations within a level depending on whether the electron is coming from an s or a p orbital, and even whether it is paired or unpaired in that orbital.

Chlorine has the electronic structure $1s^2 2s^2 2p^6 3s^2 3p_x{}^2 3p_y{}^2 3p_z{}^1$.

This graph plots the first eight ionisation energies of chlorine. The green labels show which electron is being removed for each of the ionisation energies.

If you put a ruler on the first and second points to establish the trend, you'll find that the third, fourth and fifth points lie above the value you would expect. That is because the first two electrons are coming from pairs in the 3p levels and are therefore rather easier to remove than if they were unpaired.

Again, if you put a ruler on the 3rd, 4th and 5th points to establish their trend, you'll find that the 6th and 7th points lie well above the values you would expect from a continuation of the trend. That is because the 6th and 7th electrons are coming from the 3s level - slightly closer to the nucleus and slightly less well screened.

The first 8 ionisation energies of chlorine (kJ per mole)

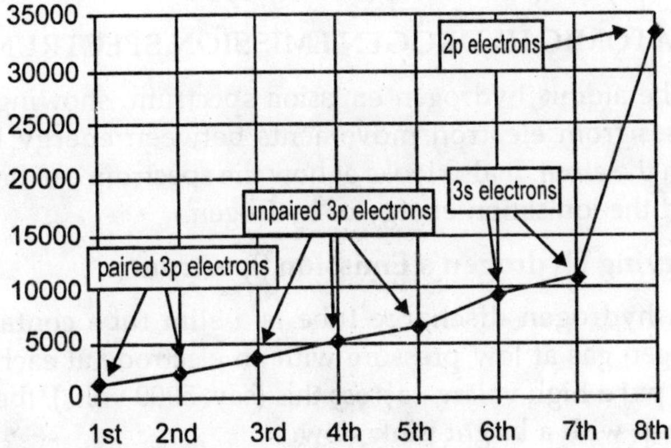

The massive jump as you break into the inner level at the 8th electron is fairly obvious.

To plot any more ionisation energies for chlorine needs a change of vertical scale. The seventeenth ionisation energy of chlorine is nearly 400,000 kJ mol^{-1}, and the vertical scale has to be squashed to accommodate this.

Log graph of all the ionisation energies of chlorine

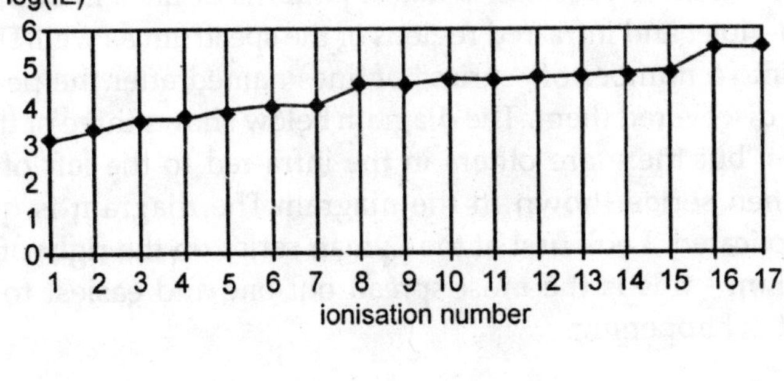

This is now a "log graph" - plotted by finding the logarithm of each ionisation energy (press the "log" button on your calculator). This doesn't simply squash the vertical scale. It distorts it as well, to such an extent that the only useful thing the graph now shows is the major jumps where the next electron to be removed comes from an inner level. The distortion is so great in the first 8 ionisation energies, for example, that the patterns shown by the previous graph are completely (and misleadingly) destroyed.

THE ATOMIC HYDROGEN EMISSION SPECTRUM

The atomic hydrogen emission spectrum, showing how it arises from electron movements between energy levels within the atom. It also looks at how the spectrum can be used to find the ionisation energy of hydrogen.

Observing Hydrogen's Emission Spectrum

A hydrogen discharge tube is a slim tube containing hydrogen gas at low pressure with an electrode at each end. If you put a high voltage across this (say, 5000 volts), the tube lights up with a bright pink glow.

If the light is passed through a prism or diffraction grating, it is split into its various colours. What you would see is a small part of the hydrogen emission spectrum. Most of the spectrum is invisible to the eye because it is either in the infra-red or the ultra-violet.

Hydrogen's Emission Spectrum

There is a lot more to the hydrogen spectrum than the three lines.It is possible to detect patterns of lines in both the ultra-violet and infra-red regions of the spectrum as well.These fall into a number of "series" of lines named after the person who discovered them. The diagram below shows three of these series, but there are others in the infra-red to the left of the Paschen series shown in the diagram.The diagram is quite complicated, Look first at the Lyman series on the right of the diagram - this is the most spread out one and easiest to see what is happening.

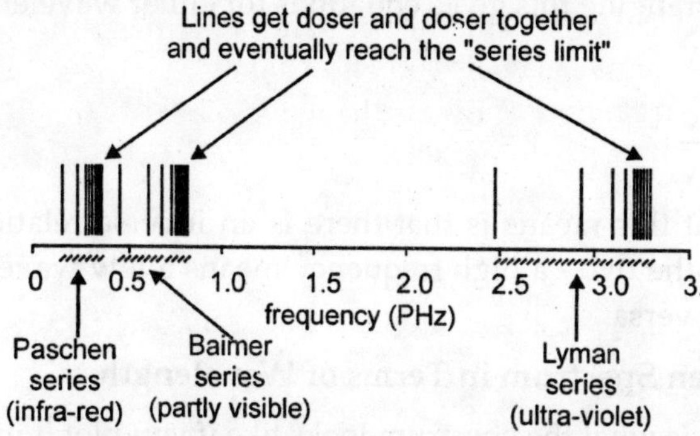

The Lyman series is a series of lines in the ultra-violet. Notice that the lines get closer and closer together as the frequency increases. Eventually, they get so close together that it becomes impossible to see them as anything other than a continuous spectrum. That's what the shaded bit on the right-hand end of the series suggests.

Then at one particular point, known as the *series limit*, the series stops.The Balmer series or the Paschen series, you will see that the pattern is just the same, but the series have become more compact. In the Balmer series.

Frequency and Wavelength

The hydrogen spectrum drawn using wavelengths of light rather than frequencies. Unfortunately, because of the mathematical relationship between the frequency of light and its wavelength, you get two completely different views of the spectrum if you plot it against frequency or against wavelength.

The Relationship between Frequency and Wavelength

The mathematical relationship is:

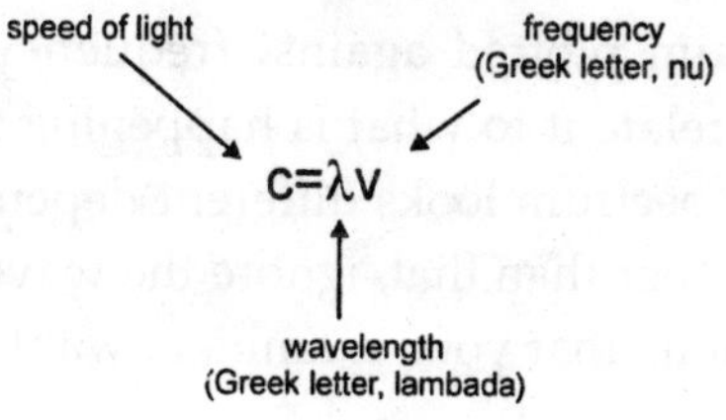

Rearranging this gives equations for either wavelength or frequency.

$$\lambda = \frac{c}{v} \qquad v = \frac{c}{\lambda}$$

What this means is that there is an inverse relationship between the two - a high frequency means a low wavelength and vice versa.

Hydrogen Spectrum in Terms of Wavelength

This is what the spectrum looks like if you plot it in terms of wavelength instead of frequency:

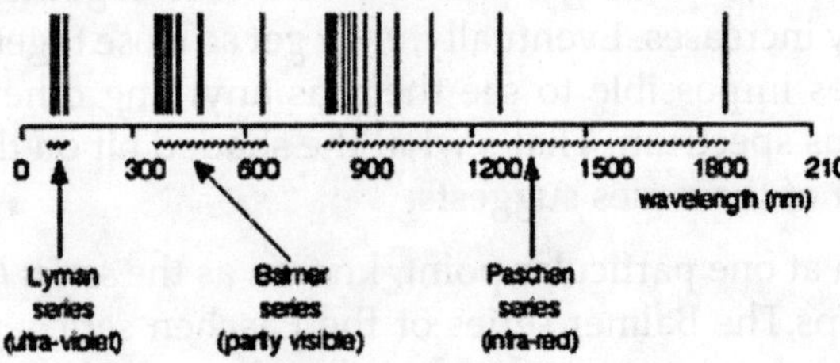

. . . and just to remind you what the spectrum in terms of frequency looks like:

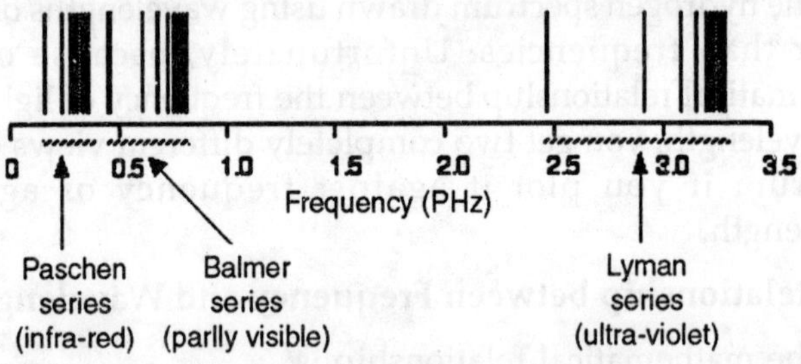

The spectrum plotted against frequency, because it is much easier to relate it to what is happening in the atom. Be aware that the spectrum looks different depending on how it is plotted, but, other than that, ignore the wavelength version unless it is obvious that your examiners want it. If you try to learn both versions, you are only going to get them muddled up!

THE BALMER AND RYDBERG EQUATIONS

By an amazing bit of mathematical insight, in 1885 Balmer came up with a simple formula for predicting the wavelength of any of the lines in what we now know as the Balmer series. Three years later, Rydberg generalised this so that it was possible to work out the wavelengths of any of the lines in the hydrogen emission spectrum.

What Rydberg came up with was:

$$\frac{1}{\lambda} = R_H\left(\frac{1}{n_1^2} - \frac{1}{n_2^2}\right)$$

R_H is a constant known as the *Rydberg constant.*

n_1 and n_2 are integers (whole numbers). n_2 has to be greater than n_1. In other words, if n_1 is, say, 2 then n_2 can be any whole number between 3 and infinity.

The various combinations of numbers that you can slot into this formula let you calculate the wavelength of any of the lines in the hydrogen emission spectrum - and there is close agreement between the wavelengths that you get using this formula and those found by analysing a real spectrum.

You can also use a modified version of the Rydberg equation to calculate the frequency of each of the lines.

$$\nu = \frac{c}{\lambda}$$

$$= c \times \frac{1}{\lambda}$$

$$\text{if } \frac{1}{\lambda} = R_H\left(\frac{1}{n_1^2} - \frac{1}{n_1^2}\right)$$

$$\nu = c.R_H\left(\frac{1}{n_1^2} - \frac{1}{n_1^2}\right)$$

The Origin of the Hydrogen Emission Spectrum

The lines in the hydrogen emission spectrum form regular patterns and can be represented by a (relatively) simple equation. Each line can be calculated from a combination of simple whole numbers.When nothing is exciting it, hydrogen's electron is in the first energy level - the level closest to the nucleus. But if you supply energy to the atom, the electron gets excited into a higher energy level - or even removed from the atom altogether.The high voltage in a discharge tube provides that energy. Hydrogen molecules are first broken up into hydrogen atoms (hence the *atomic* hydrogen emission spectrum) and electrons are then promoted into higher energy levels.Suppose a particular electron was excited into the third energy level. This would tend to lose energy again by falling back down to a lower level. It could do this in two different ways.

It could fall all the way back down to the first level again, or it could fall back to the second level - and then, in a second jump, down to the first level.

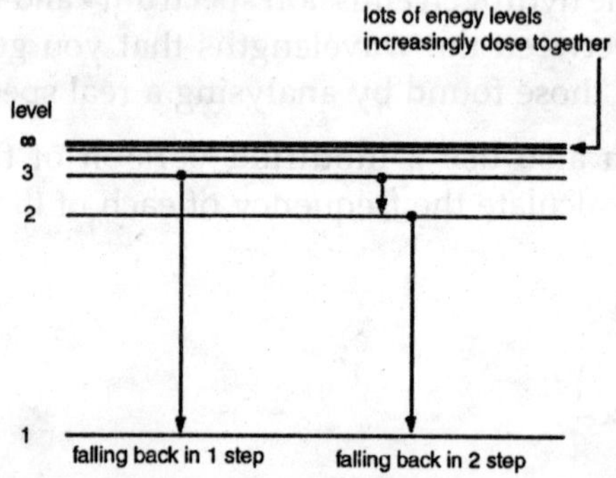

Tying Particular Electron Jumps to Individual Lines in the Spectrum

If an electron falls from the 3-level to the 2-level, it has to lose an amount of energy exactly the same as the energy gap between those two levels. That energy which the electron loses comes out as light (where "light" includes UV and IR as well as visible).

Each frequency of light is associated with a particular energy by the equation:

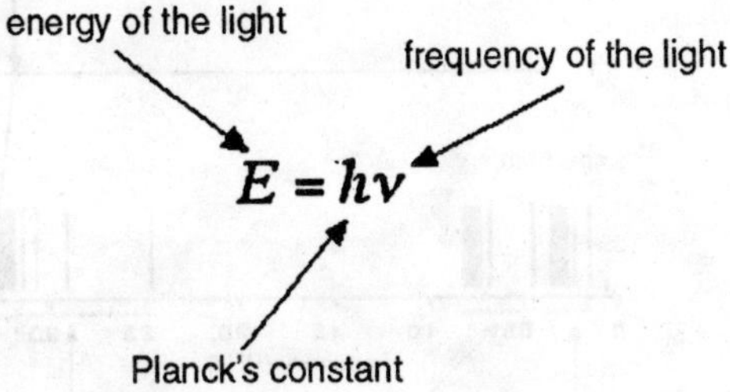

The higher the frequency, the higher the energy of the light.

If an electron falls from the 3-level to the 2-level, red light is seen. This is the origin of the red line in the hydrogen spectrum. By measuring the frequency of the red light, you can work out its energy. That energy must be exactly the same as the energy gap between the 3-level and the 2-level in the hydrogen atom.The last equation can therefore be re-written as a measure of the energy gap between two electron levels.

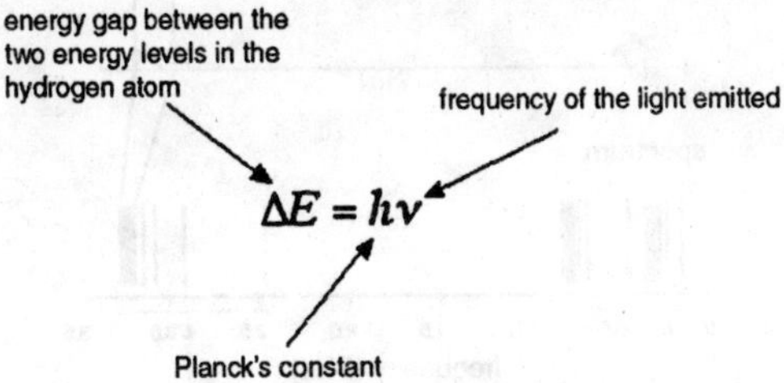

The greatest possible fall in energy will therefore produce the highest frequency line in the spectrum. The greatest fall will be from the infinity level to the 1-level. The next few diagrams are in two parts - with the energy levels at the top and the spectrum at the bottom.

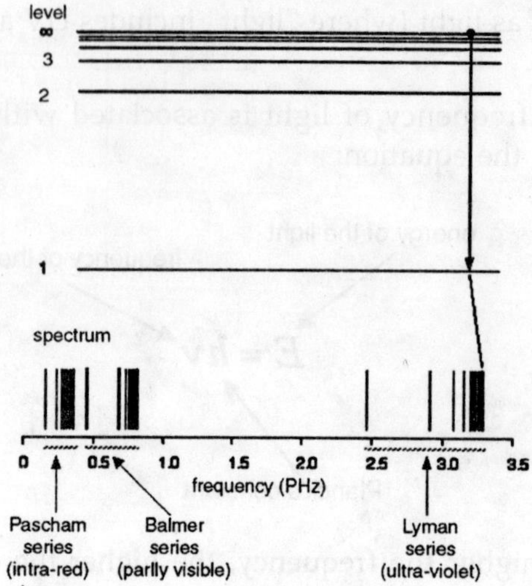

If an electron fell from the 6-level, the fall is a little bit less, and so the frequency will be a little bit lower. (Because of the scale of the diagram, it is impossible to draw in all the jumps involving all the levels between 7 and infinity!)

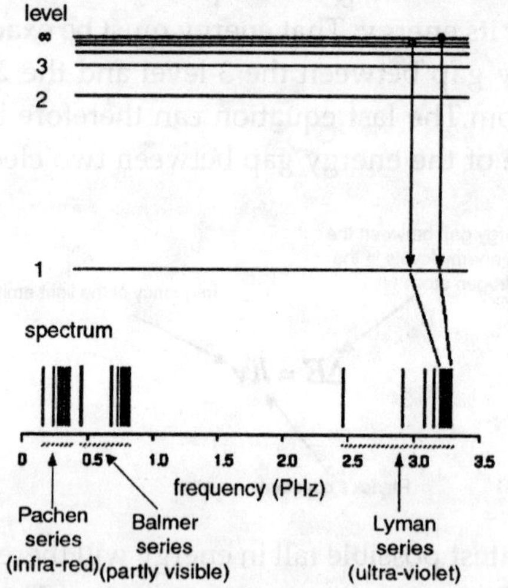

. . . and as you work your way through the other possible jumps to the 1-level, you have accounted for the whole of the Lyman series. The spacings between the lines in the spectrum reflect the way the spacings between the energy levels change.

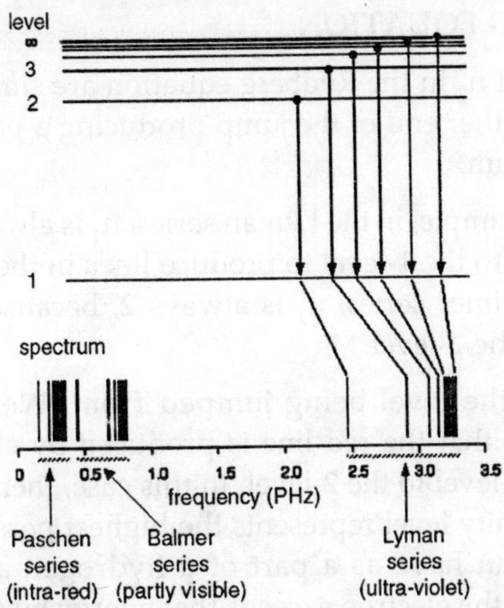

If you do the same thing for jumps down to the 2-level, you end up with the lines in the Balmer series. These energy gaps are all much smaller than in the Lyman series, and so the frequencies produced are also much lower.

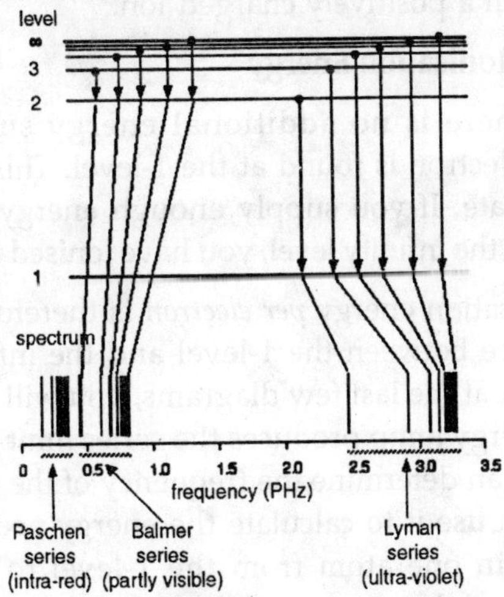

The Paschen series would be produced by jumps down to the 3-level, but the diagram is going to get very messy if I include those as well - not to mention all the other series with jumps down to the 4-level, the 5-level and so on.

RYDBERG EQUATION

n_1 and n_2 in the Rydberg equation are simply the energy levels at either end of the jump producing a particular line in the spectrum.

For example, in the Lyman series, n_1 is always 1. Electrons are falling to the 1-level to produce lines in the Lyman series. For the Balmer series, n_1 is always 2, because electrons are falling to the 2-level.

n_2 is the level being jumped from. We have already mentioned that the red line is produced by electrons falling from the 3-level to the 2-level. In this case, then, n_2 is equal to 3. The infinity level represents the highest possible energy an electron can have as a part of a hydrogen atom. So what happens if the electron exceeds that energy by even the tiniest bit?

The electron is no longer a part of the atom. The infinity level represents the point at which ionisation of the atom occurs to form a positively charged ion.

Hydrogen's Ionisation Energy

When there is no additional energy supplied to it, hydrogen's electron is found at the 1-level. This is known as its ground state. If you supply enough energy to move the electron up to the infinity level, you have ionised the hydrogen.

The ionisation energy *per electron* is therefore a measure of the distance between the 1-level and the infinity level. If you look back at the last few diagrams, you will find that that particular energy jump produces the series limit of the Lyman series.If you can determine the frequency of the Lyman series limit, you can use it to calculate the energy needed to move the electron in one atom from the 1-level to the point of ionisation. From that, you can calculate the ionisation energy per mole of atoms.

The problem is that the frequency of a series limit is quite difficult to find accurately from a spectrum because the lines are so close together in that region that the spectrum looks continuous.Here is a list of the frequencies of the seven most widely spaced lines in the Lyman series, together with the increase in frequency as you go from one to the next.

frequency ν (Hz)	increase in frequency ν (Hz)
2.467×10^{15}	
	0.457×10^{15}
2.924×10^{15}	
	0.160×10^{15}
3.084×10^{15}	
	0.074×10^{15}
3.158×10^{15}	
	0.040×10^{15}
3.198×10^{15}	
	0.025×10^{15}
3.223×10^{15}	
	0.015×10^{15}
3.238×10^{15}	

As the lines get closer together, obviously the increase in frequency gets less. At the series limit, the gap between the lines would be literally zero.

That means that if you were to plot the increases in frequency against the actual frequency, you could extrapolate (continue) the curve to the point at which the increase becomes zero. That would be the frequency of the series limit.In fact you can actually plot two graphs from the data in the table above. The frequency *difference* is related to two frequencies. For example, the figure of 0.457 is found by taking 2.467 away from 2.924. So which of these two values should you plot the 0.457 against?

It doesn't matter, as long as you are always consistent - in other words, as long as you *always* plot the difference against either the higher or the lower figure. At the point you are interested in (where the difference becomes zero), the two frequency numbers are the same.

By plotting both of the possible curves on the same graph, it makes it easier to decide exactly how to extrapolate the curves. Because these are curves, they are much more difficult to extrapolate than if they were straight lines.

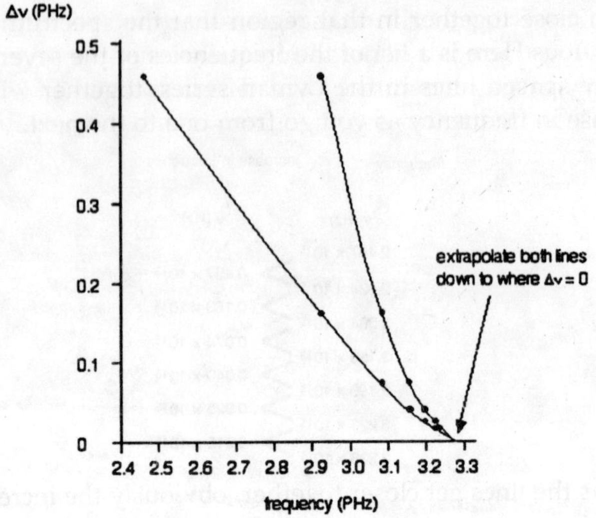

Both lines point to a series limit at about 3.28×10^{15} Hz.

So . . . now we can calculate the energy needed to remove a single electron from a hydrogen atom.

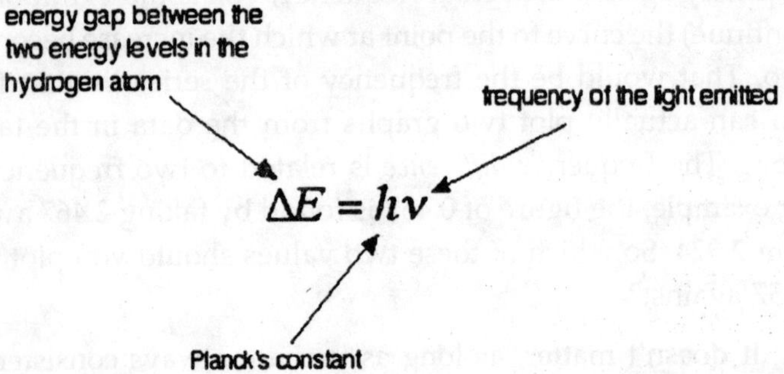

We can work out the energy gap between the ground state and the point at which the electron leaves the atom by substituting the value we've got for frequency and looking up the value of Planck's constant from a data book.

$$\Delta E = h\nu$$

$$= 6.626 \times 10^{-34} \times 3.28 \times 10^{15}$$

$$= 2.173 \times 10^{-18}\ \text{J}$$

That gives you the ionisation energy for a single atom. To find the normally quoted ionisation energy, we need to multiply this by the number of atoms in a mole of hydrogen atoms (the Avogadro constant) and then divide by 1000 to convert it into kilojoules.

$$\text{Ionisation energy} = 2.173 \times 10^{-18} \times 6.022 \times 10^{23} \times \frac{1}{1000}$$

$$= 1310\,\text{kJ mol}^{-1}$$

This compares well with the normally quoted value for hydrogen's ionisation energy of 1312 kJ mol^{-1}.

ELECTRON AFFINITY

Ionisation energies are always concerned with the formation of positive ions. Electron affinities are the negative ion equivalent, and their use is almost always confined to elements in groups 6 and 7 of the Periodic Table.The first electron affinity is the energy released when 1 mole of gaseous atoms each acquire an electron to form 1 mole of gaseous 1- ions.

This is more easily seen in symbol terms.

$$X_{(g)} + e^- \longrightarrow X^-_{(g)}$$

It is the energy released (per mole of X) when this change happens.

First electron affinities have negative values. For example, the first electron affinity of chlorine is -349 kJ mol^{-1}. By convention, the negative sign shows a release of energy.

The first electron affinities of the group 7 elements

F	-328 kJ mol^{-1}
Cl	-349 kJ mol^{-1}
Br	-324 kJ mol^{-1}
I	-295 kJ mol^{-1}

The group, first electron affinities become less (in the sense that less energy is evolved when the negative ions are formed).

Fluorine breaks that pattern, and will have to be accounted for separately. The electron affinity is a measure of the attraction between the incoming electron and the nucleus the stronger the attraction, the more energy is released. The factors which affect this attraction are exactly the same as those relating to ionisation energies - nuclear charge, distance and screening.

The increased nuclear charge as you go down the group is offset by extra screening electrons. Each outer electron in effect feels a pull of 7+ from the centre of the atom, irrespective of which element you are talking about.For example, a fluorine atom has an electronic structure of $1s^22s^22p_x^{\,2}2p_y^{\,2}2p_z^{\,1}$. It has 9 protons in the nucleus. The incoming electron enters the 2-level, and is screened from the nucleus by the two $1s^2$ electrons. It therefore feels a net attraction from the nucleus of 7+ (9 protons less the 2 screening electrons).By contrast, chlorine has the electronic structure $1s^22s^22p^63s^23p_x^{\,2}3p_y^{\,2}3p_z^{\,1}$. It has 17 protons in the nucleus.But again the incoming electron feels a net attraction from the nucleus of 7+ (17 protons less the 10 screening electrons in the first and second levels).

The over-riding factor is therefore the increased distance that the incoming electron finds itself from the nucleus as you go down the group. The greater the distance, the less the attraction and so the less energy is released as electron affinity.The incoming electron is going to be closer to the nucleus in fluorine than in any other of these elements, so you would expect a high value of electron affinity.However, because fluorine is such a small atom, you are putting the new electron into a region of space already crowded with electrons and there is a significant amount of repulsion. This repulsion lessens the attraction the incoming electron feels and so lessens the electron affinity.A similar reversal of the expected trend happens between oxygen and sulphur in Group 6. The first electron affinity of oxygen (-142 kJ mol^{-1}) is smaller than that of sulphur (-200 kJ mol^{-1}) for exactly the same reason that fluorine's is smaller than chlorine's.

COMPARING GROUP 6 AND GROUP 7 VALUES

As you might have noticed, the first electron affinity of oxygen (-142 kJ mol^{-1}) is less than that of fluorine (-328 kJ mol^{-1}). Similarly sulphur's (-200 kJ mol^{-1}) is less than chlorine's (-349 kJ mol^{-1}).It's simply that the Group 6 element has 1 less proton in the nucleus than its next door neighbour in Group 7. The amount of screening is the same in both.That means that the net pull from the nucleus is less in Group 6 than in Group 7, and so the electron affinities are less.

FIRST ELECTRON AFFINITY AND REACTIVITY

The reactivity of the elements in group 7 falls as you go down the group - fluorine is the most reactive and iodine the least.Often in their reactions these elements form their negative ions. At GCSE the impression is sometimes given that the fall in reactivity is because the incoming electron is held less strongly as you go down the group and so the negative ion is less likely to form. That explanation looks reasonable until you include fluorine!

An overall reaction will be made up of lots of different steps all involving energy changes, and you cannot safely try to explain a trend in terms of just one of those steps. Fluorine is much more reactive than chlorine (despite the lower electron affinity) because the energy released in other steps in its reactions more than makes up for the lower amount of energy released as electron affinity.

SECOND ELECTRON AFFINITY

You are only ever likely to meet this with respect to the group 6 elements oxygen and sulphur which both form 2- ions.The second electron affinity is the energy required to add an electron to each ion in 1 mole of gaseous 1- ions to produce 1 mole of gaseous 2- ions.

This is more easily seen in symbol terms.

$$X^-_{(g)} + e^- \longrightarrow X^{2-}_{(g)}$$

It is the energy needed to carry out this change per mole of X^-.

You are forcing an electron into an already negative ion. It's not going to go in willingly!

$$O_{(g)} + e^- \longrightarrow O^-_{(g)} \quad \text{1st EA} = -142 \text{ kJ mol}^{-1}$$

$$O^-_{(g)} + e^- \longrightarrow O^{2-}_{(g)} \quad \text{2nd EA} = +844 \text{ kJ mol}^{-1}$$

The positive sign shows that you have to put in energy to perform this change. The second electron affinity of oxygen is particularly high because the electron is being forced into a small, very electron-dense space.

ATOMIC AND IONIC RADIUS

Unlike a ball, an atom doesn't have a fixed radius. The radius of an atom can only be found by measuring the distance between the nuclei of two touching atoms, and then halving that distance.

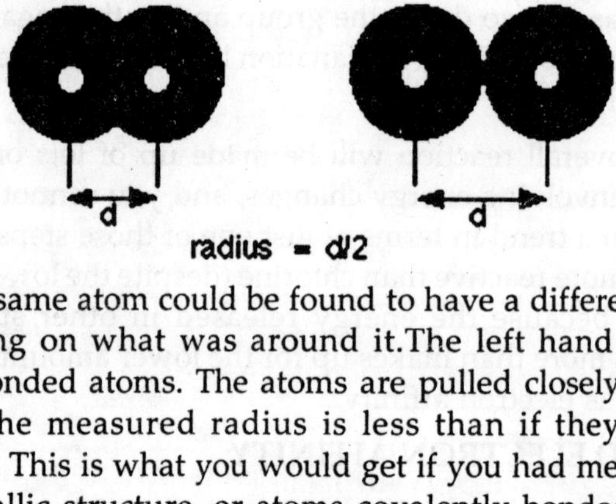

The same atom could be found to have a different radius depending on what was around it.The left hand diagram shows bonded atoms. The atoms are pulled closely together and so the measured radius is less than if they are just touching. This is what you would get if you had metal atoms in a metallic structure, or atoms covalently bonded to each other. The type of atomic radius being measured here is called the *metallic radius* or the *covalent radius* depending on the bonding.The right hand diagram shows what happens if the atoms are just touching. The attractive forces are much less, and the atoms are essentially "unsquashed". This measure of atomic radius is called the *van der Waals radius* after the weak attractions present in this situation.

The exact pattern you get depends on which measure of atomic radius you use - but the trends are still valid.The following diagram uses metallic radii for metallic elements, covalent radii for elements that form covalent bonds, and van der Waals radii for those (like the noble gases) which don't form bonds.

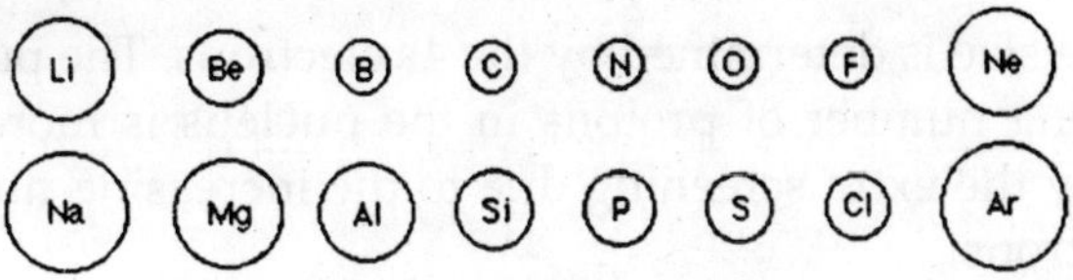

It is fairly obvious that the atoms get bigger as you go down groups. The reason is equally obvious - you are adding extra layers of electrons.

The noble gas at the end of each period. Because neon and argon don't form bonds, you can only measure their van der Waals radius - a case where the atom is pretty well "unsquashed". All the other atoms are being measured where their atomic radius is being lessened by strong attractions. You aren't comparing like with like if you include the noble gases.

Leaving the noble gases out, atoms get smaller as you go across a period.

If you think about it, the metallic or covalent radius is going to be a measure of the distance from the nucleus to the electrons which make up the bond. Look back to the left-hand side of the first diagram, and picture the bonding electrons as being half way between the two nuclei.

From lithium to fluorine, those electrons are all in the 2-level, being screened by the $1s^2$ electrons. The increasing number of protons in the nucleus as you go across the period pulls the electrons in more tightly. The amount of screening is constant for all of these elements.

In the period from sodium to chlorine, the same thing happens. The size of the atom is controlled by the 3-level bonding electrons being pulled closer to the nucleus by increasing numbers of protons - in each case, screened by the 1- and 2-level electrons.

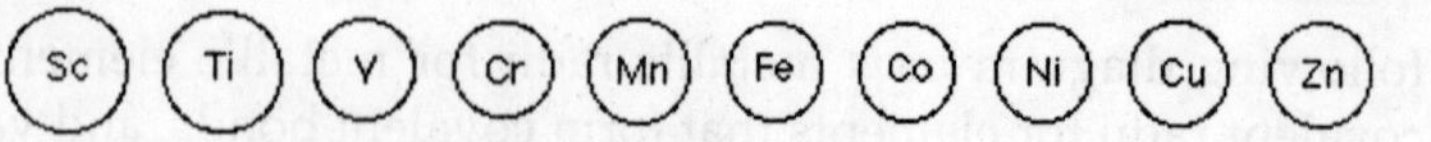

Although there is a slight contraction at the beginning of the series, the atoms are all much the same size.

The size is determined by the 4s electrons. The pull of the increasing number of protons in the nucleus is more or less offset by the extra screening due to the increasing number of 3d electrons.

IONIC RADIUS

Ions aren't the same size as the atoms they come from. Compare the sizes of sodium and chloride ions with the sizes of sodium and chlorine atoms.

Positive Ions

Positive ions are smaller than the atoms they come from. Sodium is 2,8,1; Na^+ is 2,8. You've lost a whole layer of electrons, and the remaining 10 electrons are being pulled in by the full force of 11 protons.

Negative Ions

Negative ions are bigger than the atoms they come from. Chlorine is 2,8,7; Cl^- is 2,8,8. Although the electrons are still all in the 3-level, the extra repulsion produced by the incoming electron causes the atom to expand. There are still only 17 protons, but they are now having to hold 18 electrons.

THE PHOTOELECTRIC EFFECT

Einstein successfully applied the quantum theory to the photoelectric effect - light shining onto the surface of a metal is capable of ejecting electrons from that metal. The electrons which are emitted come off at energies that depend upon the frequency of the light and (importantly) not its intensity. This can be explained if one assumes that light is composed of photons of energy hv. Einstein suggested that a photon can give up its energy to an electron in the metal. Some of this is used in work, that it needs to remove the electron from the metal, the remainder appears as kinetic energy:

$$hv = Wo + \frac{1}{2} mv^2$$

Hence, light of frequency less than hv will be incapable of ejecting an electron. This shows the dependence of the photoelectric effect on the frequency of light rather than its intensity.

Excitation of atoms - electronic spectra Niels Bohr applied the quantum theory to the phenomenon of atomic spectra. When atoms are subjected to high temperatures or electric discharges, they emit electromagnetic radiation which consists of characteristic frequencies. By subjecting the atoms of a given element to high temperatures or passing an electric discharge through a gas composed of these atoms, the atoms become excited. Electrons within the atoms are raised to higher energy levels. Once at these higher energy levels, they can return to their unexcited, or ground, state by losing this increased energy. This can be achieved by emitting electromagnetic radiation. The frequency of the emitted radiation will be determined by the energy "jump" made by the electron. The electron could return to the ground state in a single stage or it could return by a series of jumps. Rather like being at the top of a staircase and either jumping right to the foot of the stairs or by taking a series of steps.Each jump will be accompanied by a release of radiation, each being characteristic of the magnitude of the jump. Because there are a number of possible routes back to the ground state what we observe is a series of lines, each line corresponding to a jump from a higher energy level to a lower energy level. In the case of hydrogen the atomic spectra consists of a series of emission lines which have been named after their discoverers - for example the Pfund, Brackett, Paschen and Balmer series.Bohr applied quantum theory to the atomic spectrum of hydrogen and postulated that;electrons could remain in a particular orbit without emitting radiation the angular momentum of electrons in an orbit is quantized These two ideas, at conflict at that time with conventional physics, enabled Bohr to calculate an expression for the energy and radius of an atomic orbital.Although the Bohr theory gives good agreement for hydrogen and hydrogen like (He^+, Li^{2+}) it is not satisfactory for other atoms. On the basis of these ideas, Bohr and later Sommerfeld proposed models of the atom in which electrons had certain allowable energies orbiting the nucleus in circular or elliptical paths.Before going on to develop the ideas of the quantum theory, two important properties of the electron need to be discussed; the dual nature of the electron the Heisenberg Uncertainty Principle when applied to the electron

In the early part of this century (L. V. de Broglie), it was shown that matter has associated wave characteristics and this effect should be particularly marked for a particle of such a small mass as the electron. In other words, the electron can behave as both a particle and a wave.

de Broglie proposed the *wave - matter relationship*

lambda = h / mv

Experimental confirmation of the dual nature of the electron was provided by Davidson and Germer who were able to show diffraction of electrons i.e. wavelike properties.

HEISENBERG UNCERTAINTY PRINCIPLE

He was one of the greatest physicists of the twentieth century. He is best known as a founder of quantum mechanics, the new physics of the atomic world, and especially for the uncertainty principle in quantum theory.

Uncertainty Principle, also called Heisenberg Uncertainty Principle, or Indeterminacy Principle, that the position and the velocity of an object cannot both be measured exactly, at the same time, even in theory. The very concepts of exact position and exact velocity together, in fact, have no meaning in nature.

Uncertainty principle: A principle, central to quantum mechanics, which states that two complementary parameters (such as position and momentum, energy and time, or angular momentum and angular displacement) cannot both be known to infinite accuracy; the more you know about one, the less you know about the other.

It can be illustrated in a fairly clear way as it relates to position vs. momentum: To see something (let's say an electron), we have to fire photons at it; they bounce off and come back to us, so we can "see" it. If you choose low-frequency photons, with a low energy, they do not impart much momentum to the electron, but they give you a very fuzzy picture, so you have a higher uncertainty in position so that you can have a higher certainty in momentum. On the other hand, if you were to fire very high-energy photons (*x*-rays or gammas) at the electron, they would give you a very clear picture of where the electron is (higher certainty in position), but would impart a great deal of momentum to the electron (higher uncertainty in momentum).

The more precisely the position is determined, the less precisely the momentum is known in this instant, and vice versa. —Heisenberg, uncertainty

Ordinary experience provides no clue of this principle. It is easy to measure both the position and the velocity of, say, an automobile, because the uncertainties implied by this principle for ordinary objects are too small to be observed. The complete rule stipulates that the product of the uncertainties in position and velocity is equal to or greater than a tiny physical quantity, or constant (about 10-34 joule-second, the value of the quantity h (where h is Planck's constant). Only for the exceedingly small masses of atoms and subatomic particles does the product of the uncertainties become significant.

Any attempt to measure precisely the velocity of a subatomic particle, such as an electron, will knock it about in an unpredictable way, so that a simultaneous measurement of its position has no validity. This result has nothing to do with inadequacies in the measuring instruments, the technique, or the observer; it arises out of the intimate connection in nature between particles and waves in the realm of subatomic dimensions.

Every particle has a wave associated with it; each particle actually exhibits wavelike behaviour. The particle is most likely to be found in those places where the undulations of the wave are greatest, or most intense. The more intense the undulations of the associated wave become, however, the more ill defined becomes the wavelength, which in turn determines the momentum of the particle. So a strictly localized wave has an indeterminate wavelength; its associated particle, while having a definite position, has no certain velocity. A particle wave having a well-defined wavelength, on the other hand, is spread out; the associated particle, while having a rather precise velocity, may be almost anywhere. A quite accurate measurement of one observable involves a relatively large uncertainty in the measurement of the other.

The uncertainty principle is alternatively expressed in terms of a particle's momentum and position. The momentum of a particle is equal to the product of its mass times its velocity. Thus, the product of the uncertainties in the momentum and the position of a particle equals h/2 or more. The principle applies to other related pairs of observables, such as energy and time: the product of the uncertainty in an energy measurement and the uncertainty in the time interval during which the measurement is made also equals h/2 or more. The same relation holds, for an unstable atom or nucleus, between the uncertainty in the quantity of energy radiated and the uncertainty in the lifetime of the unstable system as it makes a transition to a more stable state.

$$v = \Delta x \Delta y \Delta z \, \Delta p_x \Delta p_y \Delta p_z = (2\pi\hbar)^3$$

SCHRODINGER'S WAVE EQUATION

Information about a particle in a given energy level is contained in a mathematical expression or function. This is called the wave function and is denoted by the Greek letter psi. The wave function is obtained by solving Schrodinger's Wave Equation. The square of the wave function gives the probability of finding a particle (electron) within a region of space.The Schrodinger's Wave Equation is fundamental to the development of quantum mechanics.

Acceptable solutions to the wave equation must have certain properties, and these give some insight into the significance of the wave equation.

Psi must be continuous

Psi must be finite

Psi must be single valued

the summation of psi^2 from - infinity to + infinity must = 1

Using Schrodingers equation we can calculate the probability density of the electron - i.e. the probability that the electron exists in a given volume of space. For example we can calculate this as a function of distance from the nucleus and we can define the "shape" of an orbital..Before going onto look at where we might "place" a particular electron we need to delve a little deeper into quantum mechanics.There were acceptable solutions to the wave equation. To obtain such solutions, it is necessary to use four quantum numbers to describe the orbital and the electron within that orbital. By defining the values of each of these four quantum numbers we can identify both the electron and the "orbit" of that electron. In some ways this is rather like giving a map reference for a person and indicating which way that person is facing.

Chapter 5

Chemical Bonding

TYPES OF BONDING

Chemical bonds form to lower the energy of the system, the components of the system become more stable through the formation of bonds. Everything wants to be more stable - its easy to lie down than stand up, bonding is Nature's way of allowing the elements to lie down.There are several types of chemical bond. Initially we will concern ourselves with three basic types, ionic bonding, which involves the transfer of electron(s), covalent bonding, which involves sharing of electrons and metallic bonding, which in some ways can be considered as a combination of both.If we examine the periodic table, we find that the elements in Group VIII (or 18), helium, neon, argon and so on, are particularly stable, so much so that they were once labeled the "inert gases". We now know that these elements are not inert, indeed xenon forms a range of compounds, but, nevertheless, they are very stable (although now we refer to these elements as the noble gases). This stability is the result of their electronic configuration, they have a full valence shell of electrons (ns^2, np^6) and this imparts stability. G. N. Lewis suggested that bonds (covalent) formed to enable elements to attain this "noble gas configuration". We can extend this idea to ionic compounds, in a compound such as sodium chloride, one element loses electron to gain this stable electronic configuration whilst the other gains electron to achieve the same result.

In each case, sharing or transfer of electrons, results in a more stable system. So can we predict the type of bonds which will form between different elements in the periodic table ? The answer is yes, The elements to the left of the periodic table (Groups I and II) can achieve the noble gas electronic configuration by losing electron(s). Ionization enthalpies show that lose of these is relatively easy (ionization enthalpies are fairly low for the elements in these groups). These elements are termed "*electropositive*". The opposite is true of the elements to the right of the periodic table. Elements in groups VII and VIII (17, 18) can most easily attain the noble gas configuration by gaining electron(s). These are the *electronegative* elements. (A measure of their ability to accept electrons can be seen in the *electron affinity* values for these elements).

This information allows us to predict that compounds formed by the combination of an electropositive element with an electronegative element will involve the transfer of electron(s) from the electropositive element to the electronegative element and hence ionic bonding will occur. The degree of ionic bonding will depend upon how extreme the differences are (in terms of electropositive / electronegative character) between the particular elements.

Electropositive character increases down the group whilst electronegative character increases up a group. Combination of the most highly electropositive elements (e.g. Cs, Fr) with the most electronegative element (F) will result in the "most ionic" compounds, other combinations (such as Li with Br) will be much less ionic. In other words, transfer of electron(s) is never total, for example CsF is approximately 97% ionic, there is always some degree of covalency (however small).When we get combination between two electronegative elements, the result is a sharing of electrons and so covalent bonds form. This is best illustrated by the combination of two atoms of the same element (a homonuclear diatomic). Combination of two fluorine atoms to produce F_2, for example is achieved by sharing electrons, each fluorine atom donating one electron to form a two electron (two centre, two atoms) bond. The covalent bond is formed by overlap (or combination) of atomic orbitals of each element to form a molecular orbital. The formation of molecular orbitals lies at the heart of Molecular Orbital Theory and is the subject of another set of notes. In simplistic terms, we can visualise the sharing of two electrons between atoms as forming a single bond between the two atoms, joining them together. If the atoms share two electrons each (i.e. 4 in total) we have a double bond which will be stronger (and shorter) than a single bond. Likewise, sharing six electrons gives us a triple bond.For combination of two electropositive elements we have metallic bonding. Perhaps the best way to model this is with "band theory" but that is the basis of other notes (when they are written). For now, we can simply visualise this as each metal donating electron(s) to a common "sea" of electrons which are shared by all the ions within the solid.

PROPERTIES OF COMPOUNDS

We can make some general statements about the properties of compounds based on the bonding present. These are general statements and so you will find exceptions to these descriptions but they should prove useful for a basic understanding of the chemistry involved.

Ionic Compounds

Ionic compounds are generally solids at room temperature and have high melting and boiling points. They are hard but brittle solids and are poor conductors of electricity in the solid state (good conductors when molten or in solution). To explain these properties we must examine the nature of the bonding in these compound. Ionic bonds are "strong" and omnidirectional (i.e. the ions are attracted to ions of opposite charge in all directions). The ions are (to an extent) fixed in their position within the crystal lattice and have difficulty moving from these positions, being held in position by their attraction to the surrounding ions. If the ions are moved (say through physical force) this attraction is broken and alignment of like charges can occur resulting in repulsion - hence the solid is brittle.In solution (or when molten) the ions are able to move more easily and hence conduction (through the movement of these ions) is possible. As a side note, there are compounds which are essentially ionic which do conduct electricity rather well. These are called solid electrolytes.

Covalent Compounds

In general, covalent compounds have the property of being boring at least in terms of their physical properties. They represent most of that branch of chemistry known as organic chemistry, which perhaps explains a lot. I'm joking, organic chemistry is very interesting, especially if you are an organic chemist. Covalent compounds are not boring. Maybe a little dull, but definitely not boring. In general they are gases or liquids at room temperature.One or two are solids because of extended 3 dimensional networks (diamond for example) but most are not solids. There is some interesting chemistry of covalent compounds

Metals

Metals are much more interesting. They are generally solids at room temperature. They conduct electricity and heat well and are malleable. They have high tensile strength and are usually hard. To explain these properties we must examine the bonding present. We can visualise the bonding in terms of the valence electrons of the atoms being shared throughout the metal structure. The model that tends to be applied is Band Theory but for now we can imagine the metal ions held together by this "sea" of electrons. This allows the metal to be bent and distorted without the structure breaking. These electrons are also relatively free to move and this explains the high electrical conductivity of metals.

Other Types of Bonding

We have looked briefly at the three major types of bonding but there are other, in many ways, just as important types of bonding.

Dative (Covalent) Bonding

If a atom has a lone (non-bonding) pair of electrons, this can be donated to form a dative covalent bond (provided the receiving atom can accept them). Atoms (or molecules) which have such lone pairs are called Lewis Bases, atoms (or molecules) which can accept such lone pairs are Lewis Acids.

For example ammonia, NH_3. The nitrogen has a lone pair of electrons, we can represent this as $:NH_3$, ammonia is a Lewis Base. Boron trifluoride, on the other hand is a Lewis Acid, the boron has a vacant p orbital .The boron can accept the lone pair of electrons from the nitrogen and we get the adduct F_3B-NH_3.

Polar Bonds

Models for covalent and ionic bonds are idealised and that in reality we have bonds which are intermediate between these two types. If we take HCl for example, we might suppose that there is a covalent bond between the hydrogen and the chlorine. In fact the chlorine atom is more electronegative than the hydrogen atom and so attracts more of the electron density than does the hydrogen. The upshot of this is that the chlorine becomes slightly negative and the hydrogen becomes slightly positive. The bond becomes slightly "ionic" in nature and the molecule becomes polar. It is fairly simple to decide whether or not a diatomic molecule is polar, all we need do is judge whether or not there is a significant difference in the electronegativities of the two elements present. When we get multi-atom molecules the situation is a little more complex and we have to take into account how the atoms are arranged. Carbon tetrachloride (CCl_4) for example, each C-Cl bond is polar since the chlorine is more electronegative than the carbon, but the chlorine atoms are arranged symmetrically about the carbon and so overall the molecule is non-polar. Before we can predict whether or not such a molecule is polar we have to know, or be able to predict, the shape of the molecule and this topic forms another series of notes.The effect of molecule polarity upon the physical properties of a substance can be quite startling.

What do you think might happen between different molecules of a substance when that molecule is polar ? The answer lies in the attraction between non-like charges. Opposite charges will attract one another and so in polar substance molecules will associate with one another, this is called dipole-dipole interaction.

What effect do you think this will have on, say, boiling point ?

Hydrogen Bonding

One very important example of dipole-dipole interaction has been given its own name, hydrogen bonding. An every day example of this is in water. The oxygen - hydrogen bond is polar, oxygen being the more electronegative element. The molecule is therefore polar. This is extenuated by the two lone pairs of electrons on the oxygen atom. One end of the molecule is partially negative whilst the two hydrogen atoms become partially positive. The molecules of water are attracted to one another, with the slightly positive hydrogens attracted to the negative "ends" of other water molecules. This intermolecular attraction is termed "hydrogen bonding", and acts almost like a glue holding the molecules of water together. In the case of water the effect on the physical properties of water are quite astounding, the boiling point of water, for example, is very much greater than would be the case if such bonding did not exist. This fact alone should make the human race grateful for hydrogen bonding since water would otherwise be a gas at room temperature.

Other Forces of Attraction

It is possible to liquefy the elements known as the noble gases.cool anything and it will liquefy. But what holds the atoms of these elements together. There must be some form of attraction.Imagine an atom. The atom is not a static thing, electron density fluctuates around that atom. At some point in time it is possible that the electron density on one side of the atom is greater than on the other side. We have an instantaneous dipole. If the mixture is sufficiently cold the atoms are not moving quickly then an adjacent atom will experience this dipole and will itself establish a dipole in response. We now have an attraction between the atoms instantaneous dipole induced dipole interaction.Such attraction is not strong, rather it is very weak, but nevertheless it exists.Intermolecular forces of attraction, those not involving ions, are generally referred to as van der Waal's forces.

ELECTRONS AND ORBTIALS

The subatomic particle of greatest interest with respect to chemical bonding, is the electron. Electrons are 'said' to occupy orbitals in an atom. The term orbital is somewhat 'erroneous' as electrons do not 'orbit' around a nucleus. These orbitals, also, do not really exist. They are mathematical constructs which make use of a wavefunction, ψ, that describes the motion of an electron. An orbital is, more correctly, a mathematical function, $4\pi r^2\psi^2$, which describes a region of high probability in 3D space, about a nucleus, where an electron may be found. In organic chemistry one needs to be most familiar with the s- and p-type orbitals. These orbitals can be described by using quantum numbers. The orbitals for the electrons in carbon, and the associated quantum numbers.The nucleus of the carbon atom would reside at the centre of the x,y,z coordinate.

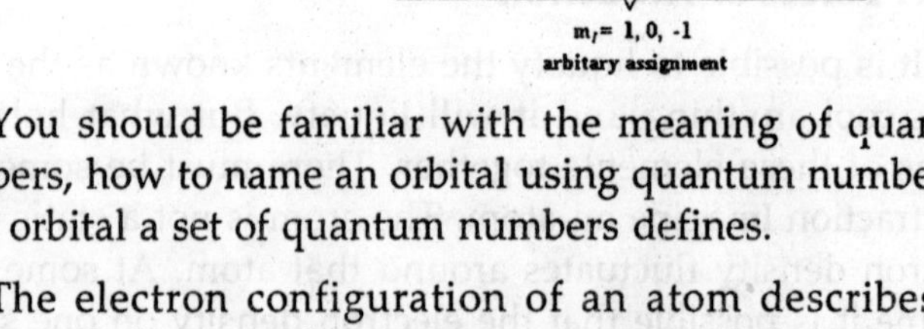

m_l = 1, 0, -1
arbitary assignment

You should be familiar with the meaning of quantum numbers, how to name an orbital using quantum numbers or what orbital a set of quantum numbers defines.

The electron configuration of an atom describes the number of electrons that atom possesses and the orbitals in which those electrons are to be found.

e.g. electron configurations of carbon

Simple	$1s^2\ 2s^2\ 2p^2$
Short form	$[He]2s^2\ 2p^2$
Orbital energy diagram	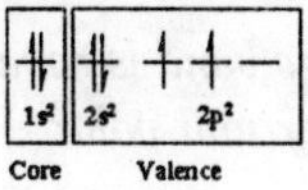

BOND TYPES

It is important to understand the structural arrangement of electrons in an atom.In everyday life, chemistry arises out of the interaction of molecules with molecules... but what is a molecule? It is a collection of atoms. When the atoms come together their electron clouds interact with one another and if this interaction is 'favourable' then bonding occurs.

One of the driving forces behind bonding is an atoms desire to obtain a stable valence electron configuration. For the atoms that are most important in organic chemistry, (N, C, O, Cl, Br, and I), this is an octet of electrons, which is similiar to the valence electron configuration of the nearest noble gas (ns^2np^6).

Ionic bonds result when electrons are tranfered between atoms.

Covalent bonds result when valence electrons are shared between atoms.

Covalent bonding present with in acetate group	$[CH_3COO]^- Na^+$	Ionic bonding between sodium and acetate group

You should be able to identify whether an ionic or covalent bond is being formed between atoms.

Most bonds within organic molecules will be covalent. The execeptions will be compounds which possess metal atoms (where the metal atoms should be treated as ions). It is important to note that the electron distribution within a covalent bond is not always symmetrical. If it is not then the bond is said to be polar as opposed to non-polar.

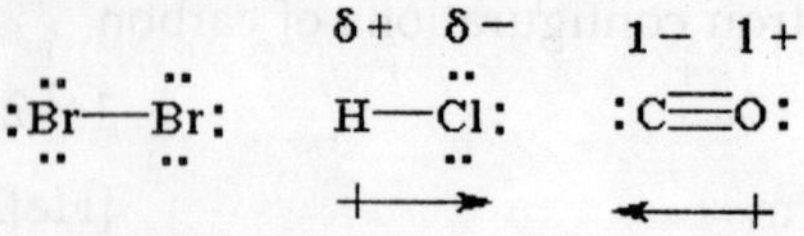

If a bond is covalent you should be able to identify whether it is polar or non-polar.

Lewis Diagrams: Lewis Diagrams give the connectivity between atoms in a molecule, they tell you nothing about the arrangement of the atoms of a molecule within 3D space.

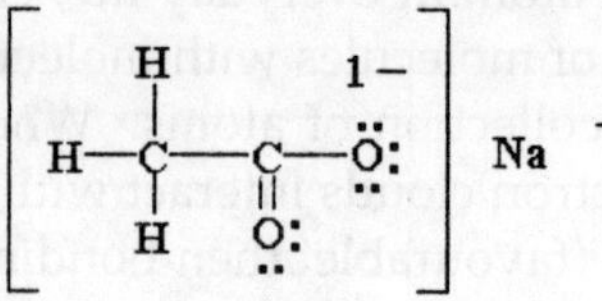

You should be able to draw a Lewis diagram for both organic and inorganic molecules as the latter are often used as reagents in organic syntheses.

In reality, electrons in a molecule are being delocalized within that molecule. Consequently, more than one Lewis diagram can be used to describe molecules which possess multiple bonds. Collectively these Lewis diagrams are then known as resonance structures and their sum is often represented as the resonance hybrid

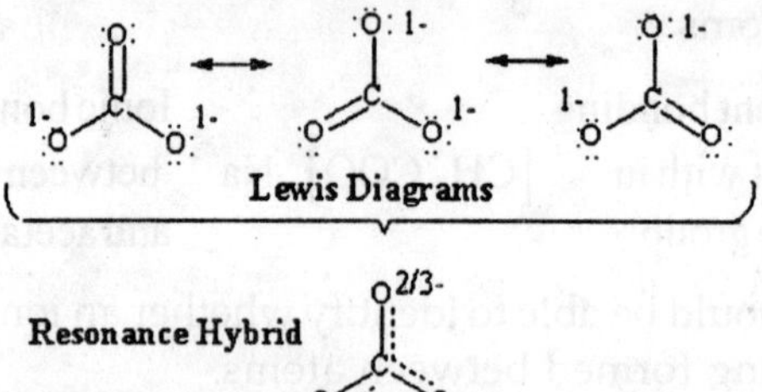

SHAPES OF MOLECULES

When determining the shape of a molecule the premise is that groups of electrons surrounding an atom are repelled as far as possible from each other to minimize the overall energy of the molecule. Groups of electrons can refer to electrons that participate in bonding or to those that are non-bonding. The number and types of groups are determined from a Lewis diagram.

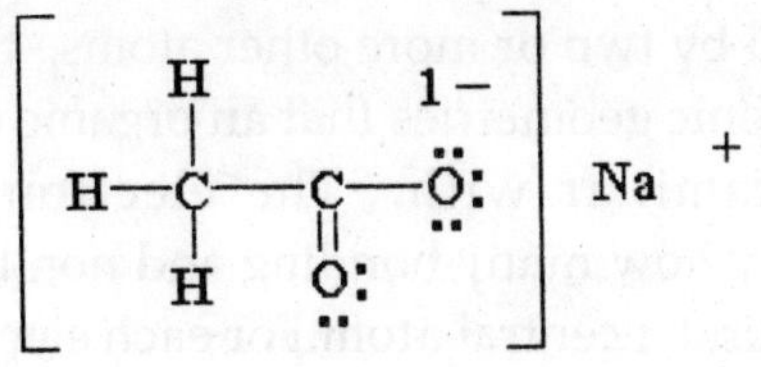

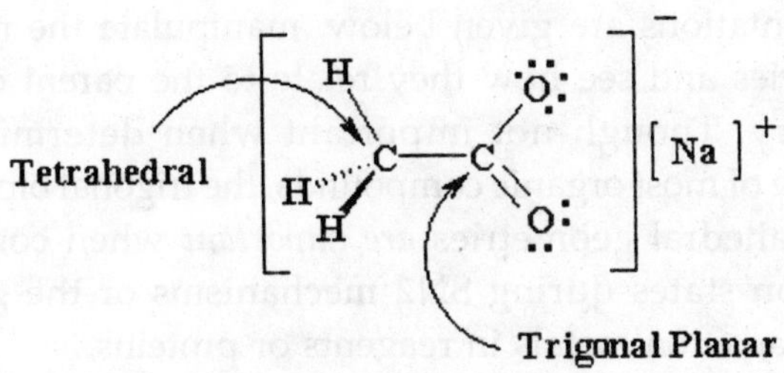

A molecule is visualized as if it were centered in the plane of the paper and, generally, so that the most number of atoms possible are in that plane. Unbroken lines are used to represent bonds in the plane of the paper, hashed wedges represent bonds going back out of the plane of paper and solid wedges represent bonds coming out of the plane of the paper. There may be several different ways that you can represent a molecule on a piece of paper. You should be able to discriminate which of those drawings *most clearly* represent the structure of a molecule.

You should be able to determine the shape of a molecule by looking at the electronic and molecular geometry about all non-terminal atoms. In doing so you must be able to present clear drawings of your conclusions..

Once the shape of a molecule is known, by referring to the sum of the bond polarities, the molecular polarity can be determined. This is important when examining the reactivity of a molecule.

You should be able to indicate a molecule's polarity.

VSEPR - Predicting the Shapes of Molecules

Looking first at simple molecules, where a central atom is surrounded by two or more other atoms, there are only six unique electronic geometries that an organic chemist needs to be readily familiar with. The electronic geometry is determined by how many bonding and non-bonding electron groups surround a central atom.For each electronic geometry, there may be a number of different molecular geometries. Representations are given below, manipulate the molecular geometries and see how they relate to the parent electronic geometry. Though not important when determining the structure of most organic compounds, the trigonal bipyramidal and octahedral geometries *are important* when considering transition states during SN2 mechanisms or the geometry about transition metals in reagents or proteins.

Electronic Geometry	Molecular Geometries
Linear	BeH_2Linear
Trigonal planar	BF_3SO_2Trigonal planarBent
Tetrahedral	CH_4 NH_3H_2STetrahedral Trigonal pyramidalBent
Trigonal bipyramidal	$PF_5TeCl_4BrF_3$ XeF_2Trigonal bipyramidal See-saw T-shaped Linear
Octahedral	$SF_6IF_5XeF_4$OctahedralSquare pyramidalSquare planar

Representing Molecular Shapes on a Piece of Paper:

Consider the simplest organic molecule, the hydrocarbon methane CH_4,

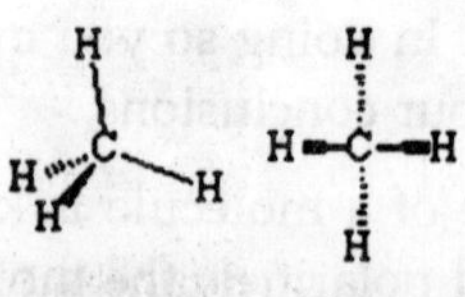

Figure I Figure II

CH_4

The trigonal bipyramidal transition state,

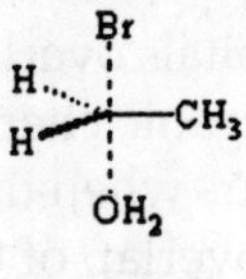
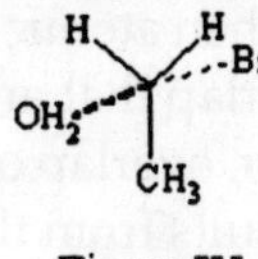
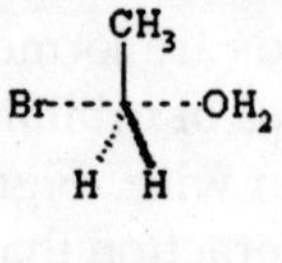
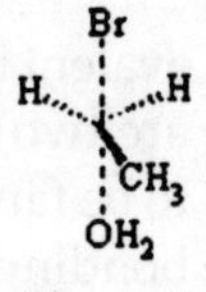

Figure III Figure IV Figure V Figure VI

C_2H_7OBr

The most desirable drawings are usually the simplest, Consider the molecule n-butane,

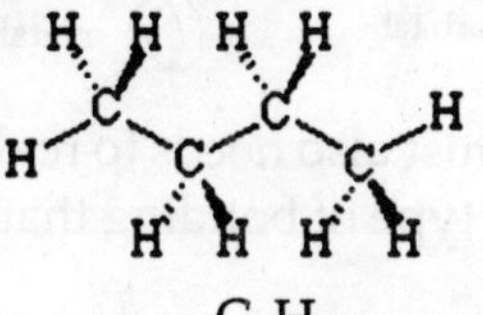

C_4H_{10}

C_4H_{10} 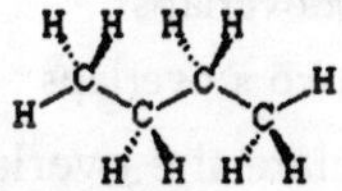

NH_3 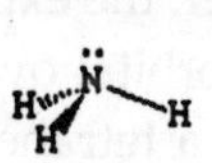 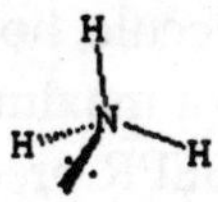

Figure X Figure XI

NH_3

Ethyl methyl ether is a structure where the opposite is true. Drawing the lone pairs in the plane of the paper would be far more confusing than if you consider them as being out of the plane.

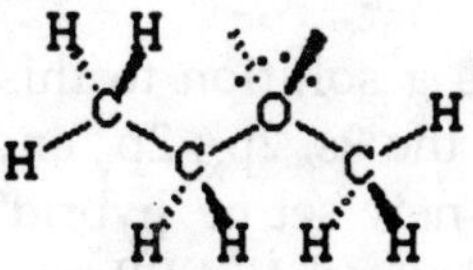

C_3H_8O

You need to learn to discriminate which drawings most clearly represent the structure of a molecule.

HYBRIDISATION:

Covalent bonds are formed when atomic orbitals overlap. There are two types of orbital overlap that an organic chemist needs to be familiar with. Sigma, σ, overlap occurs when there is one bonding interaction that results from the overlap of two orbitals. Pi, π, overlap occurs when two bonding interactions result from the overlap of orbitals.

σ overlap can result from the overlap of an s orbital with a p orbital

π overlap can result from the overlap of two p orbitals

The organic chemist also needs to realise how these orbital overlaps relate to the type of bonding that is occuring between atoms:

Single bond	σ overlap
Double bond	σ and π overlaps
Triple bond	σ and two π overlaps

If one tries to correlate the overlap of atomic orbitals to the shape of a molecule, however, the expected geometry does not correspond to a maximum orbital overlap. Take a look at methane, $CH_{4.}$ VSEPR predicts a tetrahedral geometry about the carbon atom but this is not achieved when one considers a maximum orbital overlap between four 1s orbitals of H and the 2s, $2p_x$, $2p_y$ and $2p_z$ orbitals of carbon.

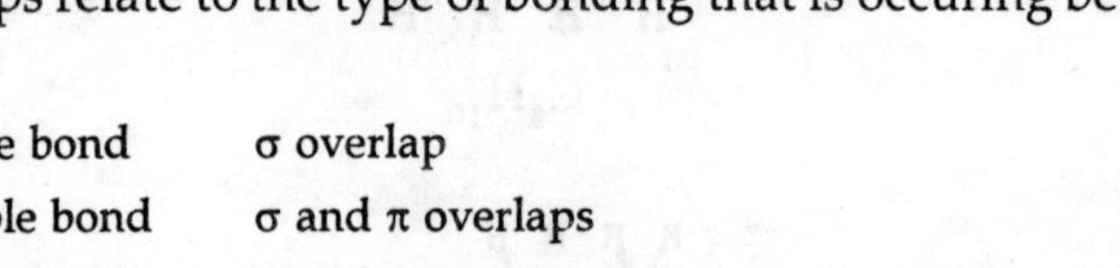

Hybridisation is a solution to this problem. It is the imaginary mixing of the 2s, $2p_x$, $2p_y$ and $2p_z$ atomic orbitals of carbon to form a new set of 'hybrid' orbitals that orient themselves in the desired VSEPR geometry. The hybrid orbitals are equivalent to one another making all orbital overlaps equivalent, therefore, all C-H bonding interactions equivalent.

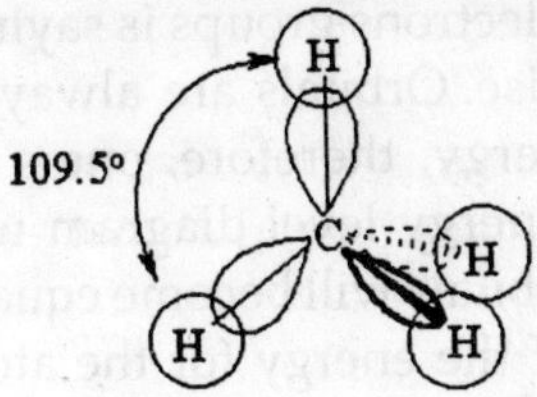

Hybrid orbitals are named by considering the type and number of atomic orbitals from which they arose. For CH_4 then the hybridisation for the carbon is sp^3. *One sees that the hybridisation of an atom can be determined very quickly by considering the number of electron groups about an atom.* Hybrid orbitals are responsible for all the s bonding overlaps in a molecule.Unhybridised orbitals are responsible for all the p bonding overlaps in a molecule.

You should be able to predict the hybridisation of all non-hydrogen and non-terminal atoms in a molecule and draw the bonding interactions in that molecule based on the hybridisation model.

Going deeper into Hybridisation:

Consider a molecule of carbon monoxide, CO. Start by visualising the imaginary mixing of the valence atomic orbitals of carbon. To do this refer to the valence orbital energy level diagram of carbon, the number of electrons is determined by taking into account any formal charges. In CO, carbon has a formal charge of one so that the electron configuration is $2s^2 2p^3$, rather than $2s^2 2p^2$.

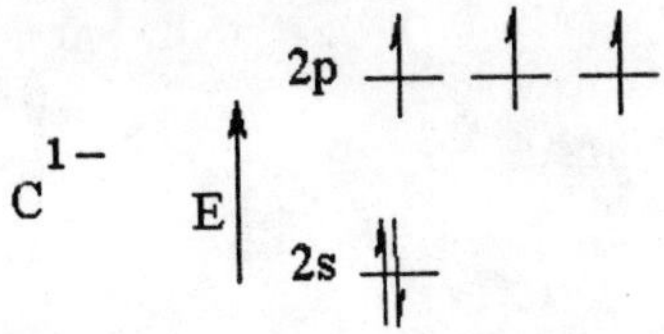

Next the number of groups about carbon for which hybridisation must occur is considered. It has two electron groups, one bonding the other a lone pair.

The number of electrons groups is saying that two atomic orbitals must hybridise. Orbitals are always hybridised from lowest to highest energy, therefore, one s and one p orbital will hybridise. The energy level diagram will change so that the two hybridised orbitals will become equal in energy, which will be an average of the energy for the atomic orbitals from which they arose, and two unhybridised p orbitals remain.

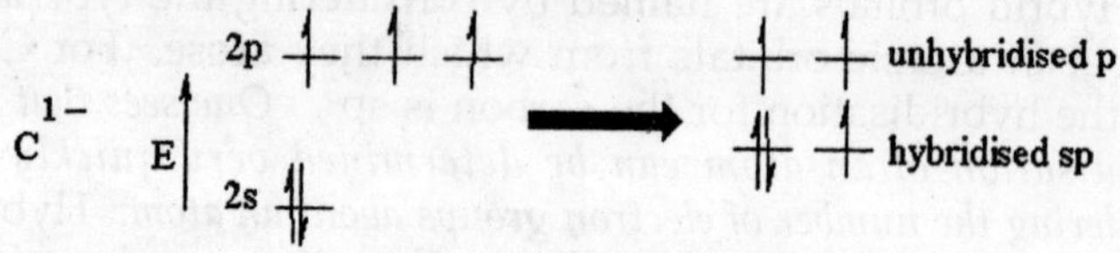

The energy level diagram for the oxygen is determined in the same way but the electron count will take into consideration the 1+ formal charge.

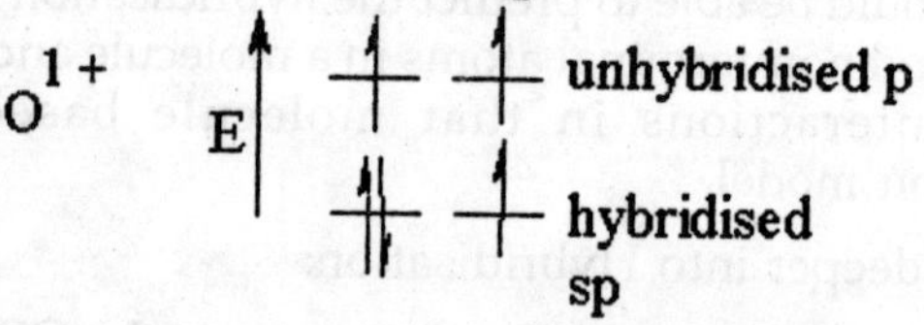

Recall that CO has a triple bond, which should be the result of one s and two p overlaps. The hybrid orbitals are involved in forming the σ overlap of the triple bond and in housing the lone pairs of electrons. The unhybridised p orbitals are involved in two p overlaps, which will be located perpendicular to one another.

Chapter 6

Oxidation and Reduction

The various definitions of oxidation and reduction in terms of the transfer of oxygen, hydrogen and electrons. It also explains the terms oxidising agent and reducing agent.

Oxidation and Reduction in Terms of Oxygen Transfer

Definitions

- Oxidation is gain of oxygen.
- Reduction is loss of oxygen.

For example, in the extraction of iron from its ore:

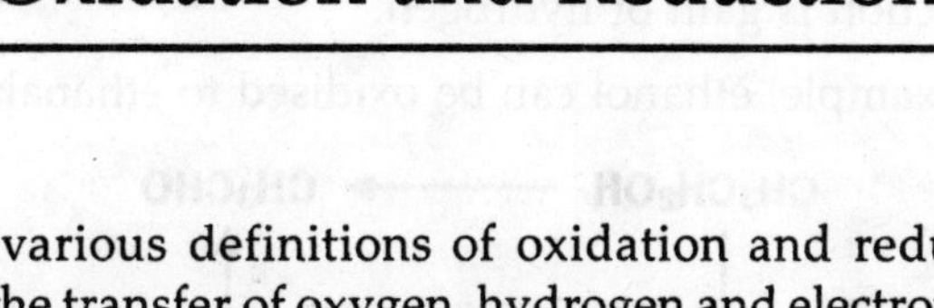

Because both *red*uction and *ox*idation are going on side-by-side, this is known as a *redox* reaction.

Oxidising and Reducing Agents

An oxidising agent is substance which oxidises something else. In the above example, the iron(III) oxide is the oxidising agent.A reducing agent reduces something else. In the equation, the carbon monoxide is the reducing agent.

Oxidising agents give oxygen to another substance.

Reducing agents remove oxygen from another substance.

Oxidation and Reduction In terms of Hydrogen Transfer

These are old definitions which aren't used very much nowadays. The most likely place you will come across them is in organic chemistry.

Definitions

Oxidation is loss of hydrogen.

Reduction is gain of hydrogen.

For example, ethanol can be oxidised to ethanal:

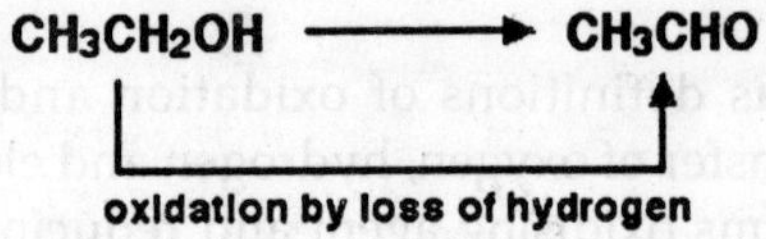

You would need to use an oxidising agent to remove the hydrogen from the ethanol. A commonly used oxidising agent is potassium dichromate(VI) solution acidified with dilute sulphuric acid.

Ethanal can also be reduced back to ethanol again by adding hydrogen to it. A possible reducing agent is sodium tetrahydridoborate, $NaBH_4$. Again the equation is too complicated to be worth bothering about at this point.

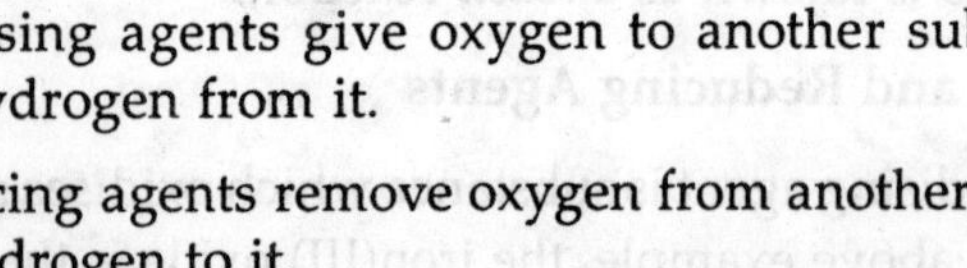

An update on oxidising and reducing agents

Oxidising agents give oxygen to another substance or remove hydrogen from it.

Reducing agents remove oxygen from another substance or give hydrogen to it.

Oxidation and Reduction in Terms of Electron Transfer

Definitions

Oxidation is loss of electrons.

Reduction is gain of electrons.

It is essential that you remember these definitions. There is a very easy way to do this. As long as you remember that you are talking about electron transfer:

OIL	RIG
oxidation is loss	reduction is gain

A simple example

The equation shows a simple redox reaction which can obviously be described in terms of oxygen transfer.

$$CuO + Mg \longrightarrow Cu + MgO$$

Copper(II) oxide and magnesium oxide are both ionic. The metals obviously aren't. If you rewrite this as an ionic equation, it turns out that the oxide ions are spectator ions and you are left with:

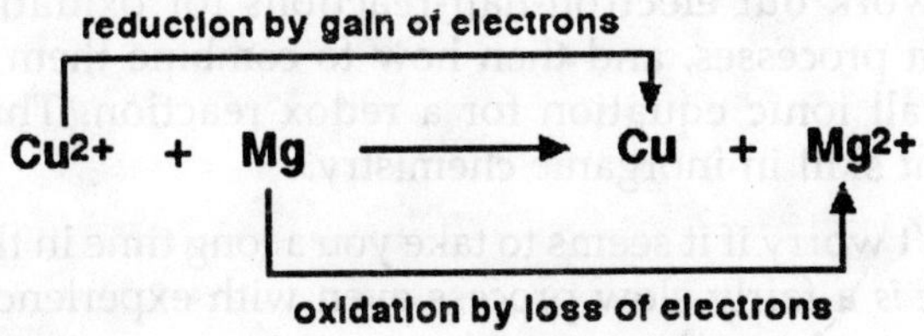

A last comment on oxidising and reducing agents

If you look at the equation above, the magnesium is reducing the copper(II) ions by giving them electrons to neutralise the charge. Magnesium is a reducing agent.Looking at it the other way round, the copper(II) ions are removing electrons from the magnesium to create the magnesium ions. The copper(II) ions are acting as an oxidising agent.This is potentially very confusing if you try to learn both what oxidation and reduction mean in terms of electron transfer, and also learn definitions of oxidising and reducing agents in the same terms.

Personally, I would recommend that you work it out if you need it. The argument (going on inside your head) would go like this if you wanted to know, for example, what an oxidising agent did in terms of electrons:

- An oxidising agent oxidises something else.
- Oxidation is loss of electrons (OIL RIG).
- That means that an oxidising agent takes electrons from that other substance.
- So an oxidising agent must gain electrons.

Or you could think it out like this:

- An oxidising agent oxidises something else.
- That means that the oxidising agent must be being reduced.
- Reduction is gain of electrons (OIL RIG).
- So an oxidising agent must gain electrons.
- Writing Ionic Equation For Redox Reaction

To work out electron-half-reactions for oxidation and reduction processes, and then how to combine them to give the overall ionic equation for a redox reaction. This is an important skill in inorganic chemistry.

Don't worry if it seems to take you a long time in the early stages. It *is* a fairly slow process even with experience. Take your time and practice as much as you can.

Electron-half-equations

When magnesium reduces hot copper(II) oxide to copper, the ionic equation for the reaction is:

$$Cu^{2+} + Mg \longrightarrow Cu + Mg^{2+}$$

You can split the ionic equation into two parts, and look at it from the point of view of the magnesium and of the copper(II) ions separately. This shows clearly that the magnesium has lost two electrons, and the copper(II) ions have gained them.

$$Mg \longrightarrow Mg^{2+} + 2e^-$$

$$Cu^{2+} + 2e^- \longrightarrow Cu$$

These two equations are described as "electron-half-equations" or "half-equations" or "ionic-half-equations" or "half-reactions" - lots of variations all meaning exactly the same thing!

Any redox reaction is made up of two half-reactions: in one of them electrons are being lost (an oxidation process) and in the other one those electrons are being gained (a reduction process).

Working out electron-half-equations and using them to build ionic equations.

In the example above, we've got at the electron-half-equations by starting from the ionic equation and extracting the individual half-reactions from it. That's doing everything entirely the wrong way round!

In reality, you almost always start from the electron-half-equations and use them to build the ionic equation.

Example 1: The reaction between chlorine and iron(II) ions

Chlorine gas oxidises iron(II) ions to iron(III) ions. In the process, the chlorine is reduced to chloride ions.

You would have to know this, or be told it by an examiner. In building equations, there is quite a lot that you can work out as you go along, but you have to have somewhere to start from!You start by writing down what you know for each of the half-reactions. In the chlorine case, you know that chlorine (as molecules) turns into chloride ions:

$$Cl_2 \longrightarrow Cl^-$$

The first thing to do is to balance the atoms that you have got as far as you possibly can:

$$Cl_2 \longrightarrow 2Cl^-$$

Always check that you have the existing atoms balanced before you do anything else. If you forget to do this, everything else that you do afterwards is a complete waste of time!

Now you have to add things to the half-equation in order to make it balance completely.

All you are allowed to add are:

• Electrons • Water

Hydrogen ions (unless the reaction is being done under alkaline conditions - in which case, you can add hydroxide ions instead)

In the chlorine case, all that is wrong with the existing equation that we've produced so far is that the charges don't balance. The left-hand side of the equation has no charge, but the right-hand side carries 2 negative charges.

That's easily put right by adding two electrons to the left-hand side. The final version of the half-reaction is:

$$Cl_2 + 2e^- \longrightarrow 2Cl^-$$

Now you repeat this for the iron(II) ions. You know (or are told) that they are oxidised to iron(III) ions. Write this down:

$$Fe^{2+} \longrightarrow Fe^{3+}$$

The atoms balance, but the charges don't. There are 3 positive charges on the right-hand side, but only 2 on the left.

You need to reduce the number of positive charges on the right-hand side. That's easily done by adding an electron to that side:

$$Fe^{2+} \longrightarrow Fe^{3+} + e^-$$

Combining the half-reactions to make the ionic equation for the reaction

What we've got at the moment is this:

This reaction needs 2 electrons . . .

$$Cl_2 + 2e^- \longrightarrow 2Cl^-$$

$$Fe^{2+} \longrightarrow Fe^{3+} + e^-$$

. . . but this is only producing 1 electron

It is obvious that the iron reaction will have to happen twice for every chlorine molecule that reacts. Allow for that, and then add the two half-equations together.

$$1 \times (Cl_2 + 2e^- \longrightarrow 2Cl^-)$$

$$2 \times (Fe^{2+} \longrightarrow Fe^{3+} + e^-)$$

$$Cl_2 + 2Fe^{2+} \longrightarrow 2Cl^- + 2Fe^{3+}$$

But don't stop there!! Check that everything balances - atoms and charges. It is very easy to make small mistakes, especially if you are trying to multiply and add up more complicated equations.

The reaction between hydrogen peroxide and manganate(VII) ions

The first example was a simple bit of chemistry which you may well have come across. The technique works just as well for more complicated (and perhaps unfamiliar) chemistry.

Manganate(VII) ions, MnO_4^-, oxidise hydrogen peroxide, H_2O_2, to oxygen gas. The reaction is done with potassium manganate(VII) solution and hydrogen peroxide solution acidified with dilute sulphuric acid.

During the reaction, the manganate(VII) ions are reduced to manganese(II) ions.

Let's start with the hydrogen peroxide half-equation. What we know is:

$$H_2O_2 \longrightarrow O_2$$

The oxygen is already balanced. What about the hydrogen?

All you are allowed to add to this equation are water, hydrogen ions and electrons. If you add water to supply the extra hydrogen atoms needed on the right-hand side, you will mess up the oxygens again - that's obviously wrong!

Add two hydrogen ions to the right-hand side.

$$H_2O_2 \longrightarrow O_2 + 2H^+$$

Now all you need to do is balance the charges. You would have to add 2 electrons to the right-hand side to make the overall charge on both sides zero.

$$H_2O_2 \longrightarrow O_2 + 2H^+ + 2e^-$$

Now for the manganate(VII) half-equation:

You know (or are told) that the manganate(VII) ions turn into manganese(II) ions. Write that down.

$$MnO_4^- \longrightarrow Mn^{2+}$$

The manganese balances, but you need four oxygens on the right-hand side. These can only come from water - that's the only oxygen-containing thing you are allowed to write into one of these equations in acid conditions.

$$MnO_4^- \longrightarrow Mn^{2+} + 4H_2O$$

By doing this, we've introduced some hydrogens. To balance these, you will need 8 hydrogen ions on the left-hand side.

$$MnO_4^- + 8H^+ \longrightarrow Mn^{2+} + 4H_2O$$

Now that all the atoms are balanced, all you need to do is balance the charges. At the moment there are a net 7+ charges on the left-hand side (1- and 8+), but only 2+ on the right. Add 5 electrons to the left-hand side to reduce the 7+ to 2+.

$$MnO_4^- + 8H^+ + 5e^- \longrightarrow Mn^{2+} + 4H_2O$$

This is the typical sort of half-equation which you will have to be able to work out. The sequence is usually:

- Balance the atoms apart from oxygen and hydrogen.
- Balance the oxygens by adding water molecules.
- Balance the hydrogens by adding hydrogen ions.
- Balance the charges by adding electrons.
- Combining the half-reactions to make the ionic equation for the reaction

The two half-equations we've produced are:

This reaction needs 5 electrons . . .

$$MnO_4^- + 8H^+ + 5e^- \longrightarrow Mn^{2+} + 4H_2O$$

$$H_2O_2 \longrightarrow O_2 + 2H^+ + 2e^-$$

. . . but this is only producing 2 electrons

You have to multiply the equations so that the same number of electrons are involved in both. In this case, everything would work out well if you transferred 10 electrons.

$$2 \times (MnO_4^- + 8H^+ + 5e^- \longrightarrow Mn^{2+} + 4H_2O)$$

$$5 \times (H_2O_2 \longrightarrow O_2 + 2H^+ + 2e^-)$$

$$2MnO_4^- + 16H^+ + 5H_2O_2 \longrightarrow 2Mn^{2+} + 8H_2O + 5O_2 + 10H^+$$

But this time, you haven't quite finished. During the checking of the balancing, you should notice that there are hydrogen ions on both sides of the equation:

hydrogen ions on both sides of the equation

$$2MnO_4^- + 16H^+ + 5H_2O_2 \longrightarrow 2Mn^{2+} + 8H_2O + 5O_2 + 10H^+$$

You can simplify this down by subtracting 10 hydrogen ions from both sides to leave the final version of the ionic equation - but don't forget to check the balancing of the atoms and charges!

$$2MnO_4^- + 6H^+ + 5H_2O_2 \longrightarrow 2Mn^{2+} + 8H_2O + 5O_2$$

You will often find that hydrogen ions or water molecules appear on both sides of the ionic equation in complicated cases built up in this way. Always check, and then simplify where possible.

Example 3: The oxidation of ethanol by acidified potassium dichromate(VI)

This technique can be used just as well in examples involving organic chemicals. Potassium dichromate(VI) solution acidified with dilute sulphuric acid is used to oxidise ethanol, CH_3CH_2OH, to ethanoic acid, CH_3COOH.

The oxidising agent is the dichromate(VI) ion, $Cr_2O_7^{2-}$. This is reduced to chromium(III) ions, Cr^{3+}.

We'll do the ethanol to ethanoic acid half-equation first. Using the same stages as before, start by writing down what you know:

$$CH_3CH_2OH \longrightarrow CH_3COOH$$

Balance the oxygens by adding a water molecule to the left-hand side:

$$CH_3CH_2OH + H_2O \longrightarrow CH_3COOH$$

Add hydrogen ions to the right-hand side to balance the hydrogens:

$$CH_3CH_2OH + H_2O \longrightarrow CH_3COOH + 4H^+$$

And finally balance the charges by adding 4 electrons to the right-hand side to give an overall zero charge on each side:

$$CH_3CH_2OH + H_2O \longrightarrow CH_3COOH + 4H^+ + 4e^-$$

The dichromate(VI) half-equation contains a trap which lots of people fall into!

Start by writing down what you know:

$$Cr_2O_7^{2-} \longrightarrow Cr^{3+}$$

What people often forget to do at this stage is to balance the chromiums. If you don't do that, you are doomed to getting the wrong answer at the end of the process! When you come to balance the charges you will have to write in the wrong number of electrons - which means that your multiplying factors will be wrong when you come to add the half-equations . . . A complete waste of time!

$$Cr_2O_7^{2-} \longrightarrow 2Cr^{3+}$$

Now balance the oxygens by adding water molecules . . .

$$Cr_2O_7^{2-} \longrightarrow 2Cr^{3+} + 7H_2O$$

. . . and the hydrogens by adding hydrogen ions:

$$Cr_2O_7^{2-} + 14H^+ \longrightarrow 2Cr^{3+} + 7H_2O$$

Now all that needs balancing is the charges. Add 6 electrons to the left-hand side to give a net 6+ on each side.

$$Cr_2O_7^{2-} + 14H^+ + 6e^- \longrightarrow 2Cr^{3+} + 7H_2O$$

Combining the half-reactions to make the ionic equation for the reaction

What we have so far is:

$$CH_3CH_2OH + H_2O \longrightarrow CH_3COOH + 4H^+ + 4e^-$$

$$Cr_2O_7^{2-} + 14H^+ + 6e^- \longrightarrow 2Cr^{3+} + 7H_2O$$

What are the multiplying factors for the equations this time? The simplest way of working this out is to find the smallest number of electrons which both 4 and 6 will divide into - in this case, 12. That means that you can multiply one equation by 3 and the other by 2.

The multiplication and addition looks like this:

$$3 \times (CH_3CH_2OH + H_2O \longrightarrow CH_3COOH + 4H^+ + 4e^-)$$

$$2 \times (Cr_2O_7^{2-} + 14H^+ + 6e^- \longrightarrow 2Cr^{3+} + 7H_2O)$$

$$3CH_3CH_2OH + 3H_2O + 2Cr_2O_7^{2-} + 28H^+ \longrightarrow 3CH_3COOH + 12H^+ + 4Cr^{3+} + 14H_2O$$

Now you will find that there are water molecules and hydrogen ions occurring on both sides of the ionic equation. You can simplify this to give the final equation:

$$3CH_3CH_2OH + 2Cr_2O_7^{2-} + 16H^+ \longrightarrow 3CH_3COOH + 4Cr^{3+} + 11H_2O$$

Reactions done under alkaline conditions

Writing Ionic Equations for Redox Reactions Done under Alkaline Conditions.

To work out electron-half-reactions for oxidation and reduction processes which are carried out under alkaline conditions, and then how to combine them to give the overall ionic equation for the redox reaction. Combining them is easy; working them out may be more difficult than under acidic conditions.

When you are trying to balance electron-half-equations, you are only allowed to add:

- Electrons
- Water

hydrogen ions (unless the reaction is being done under alkaline conditions - in which case, you can add hydroxide ions instead)

If you are working under *acidic or neutral* conditions, the sequence is usually:

- Balance the atoms apart from oxygen and hydrogen.
- Balance the oxygens by adding water molecules.
- Balance the hydrogens by adding hydrogen ions.
- Balance the charges by adding electrons.

The whole process is fairly automatic, and provided you take care, there isn't much to go wrong.

The problem is that the water and the hydroxide ions that you add to balance the equations under alkaline conditions contain both hydrogen and oxygen.To balance the oxygens, you could in principle add either H_2O or OH^- to the equation. The same thing is true for balancing the hydrogens. How do you know what to start with?

In some cases, it is obvious how to build up the half-equation using hydroxide ions. Always check this before you get involved in anything more difficult.

If it isn't immediately obvious, work out the electron-half equation *as if it were being done under acidic conditions* in other words by writing in water molecules, hydrogen ions and electrons.

Once you have got a balanced half-equation, you then convert it to alkaline conditions.

Four examples:

Don't worry if the chemistry in these examples is unfamiliar to you. It doesn't matter in the slightest. All that matters is how you work out the equations.

The oxidation of cobalt(II) to cobalt(III) by hydrogen peroxide

If you add an excess of ammonia solution to a solution containing cobalt(II) ions, you get a complex ion formed called the hexaamminecobalt(II) ion, $Co(NH_3)_6^{2+}$. This is oxidised rapidly by hydrogen peroxide solution to the hexaamminecobalt(III) ion, $Co(NH_3)_6^{3+}$.

Ammonia solution is, of course, alkaline.

The half-equation for the cobalt reaction is easy. Start by writing down what you know (or are told):

$$Co(NH_3)_6^{2+} \longrightarrow Co(NH_3)_6^{3+}$$

Everything balances apart from the charges. Add an electron to the right-hand side to give both sides an overall charge of 2+.

$$Co(NH_3)_6^{2+} \longrightarrow Co(NH_3)_6^{3+} + e^-$$

The hydrogen peroxide half-equation isn't very difficult either, except that you aren't told what is formed and so have to make a guess. It would balance very nicely if you ended up with 2 hydroxide ions on the right-hand side.

This is a good example of a case where it is fairly obvious where to put hydroxide ions.

$$H_2O_2 \longrightarrow 2OH^-$$

You would then just have to add 2 electrons to the left-hand side to balance the charges.

$$H_2O_2 + 2e^- \longrightarrow 2OH^-$$

Combining the half-reactions to make the ionic equation for the reaction

What we have so far is:

$$Co(NH_3)_6^{2+} \longrightarrow Co(NH_3)_6^{3+} + e^-$$

$$H_2O_2 + 2e^- \longrightarrow 2OH^-$$

The multiplication and addition looks like this:

$$2 \times (Co(NH_3)_6^{2+} \longrightarrow Co(NH_3)_6^{3+} + e^-)$$

$$1 \times (H_2O_2 + 2e^- \longrightarrow 2OH^-)$$

$$2Co(NH_3)_6^{2+} + H_2O_2 \longrightarrow 2Co(NH_3)_6^{3+} + 2OH^-$$

And that's it - an easy example!

The oxidation of iron(II) hydroxide by the air

If you add sodium hydroxide solution to a solution of an iron(II) compound you get a green precipitate of iron(II) hydroxide. This is quite quickly oxidised by oxygen in the air to an orange-brown precipitate of iron(III) hydroxide.

The half-equation for the iron(II) hydroxide is straightforward. Start with what you know:

$$Fe(OH)_2 \longrightarrow Fe(OH)_3$$

You obviously need another hydroxide ion on the left-hand side. This is even more straightforward than the previous example.

$$Fe(OH)_2 + OH^- \longrightarrow Fe(OH)_3$$

To balance the charges, add an electron to the right-hand side.

$$Fe(OH)_2 + OH^- \longrightarrow Fe(OH)_3 + e^-$$

The half-equation for the oxygen isn't so easy. You don't know what is being formed.

$$O_2 \longrightarrow ?$$

It isn't at all obvious whether you need to balance the oxygens by adding water molecules or hydroxide ions to the right-hand side. OK - treat it as if it were being done under acidic conditions, and the problem disappears!

In this case, you can only balance the oxygens by adding water molecules to the right-hand side.

$$O_2 \longrightarrow 2H_2O$$

Balance the hydrogens by adding hydrogen ions to the left-hand side.

$$O_2 + 4H^+ \longrightarrow 2H_2O$$

And then balance the charges by adding 4 electrons:

$$O_2 + 4H^+ + 4e^- \longrightarrow 2H_2O$$

Now you have got a perfectly balanced half-equation. The problem is, of course, that it only applies under acidic conditions. We *should* have alkaline conditions - with hydroxide ions present not hydrogen ions.

So . . . get rid of the hydrogen ions! Add enough hydroxide ions to *both* sides of the equation so that you can neutralise all the hydrogen ions. Because it is now a balanced equation, you must add the same number to *both* sides to maintain the balance.

$$O_2 + 4H^+ + 4OH^- + 4e^- \longrightarrow 2H_2O + 4OH^-$$

The hydrogen ions and hydroxide ions on the left-hand side would turn into 4 water molecules:

$$O_2 + 4H_2O + 4e^- \longrightarrow 2H_2O + 4OH^-$$

Finally, there are water molecules on both sides of the equation. Cancel out any which aren't changed.

$$O_2 + 2H_2O + 4e^- \longrightarrow 4OH^-$$

This has all been a bit tedious - although you haven't actually had to think very much! Don't forget to re-check that everything balances now that you have finished.

Combining the half-reactions to make the ionic equation for the reaction

From here on it's all back to the usual routine. We've worked out the two half-equations:

$$Fe(OH)_2 + OH^- \longrightarrow Fe(OH)_3 + e^-$$

$$O_2 + 2H_2O + 4e^- \longrightarrow 4OH^-$$

The iron equation will have to happen 4 times to supply enough electrons to the oxygen.

$$4 \times (Fe(OH)_2 + OH^- \longrightarrow Fe(OH)_3 + e^-)$$

$$1 \times (O_2 + 2H_2O + 4e^- \longrightarrow 4OH^-)$$

$$4Fe(OH)_2 + 4OH^- + O_2 + 2H_2O \longrightarrow 4Fe(OH)_3 + 4OH^-$$

Notice that the hydroxide ions on each side cancel out. (Perhaps to your surprise - certainly to mine when I worked this out!)

$$4Fe(OH)_2 + O_2 + 2H_2O \longrightarrow 4Fe(OH)_3$$

The reduction of manganate(VII) ions to manganate(VI) ions by hydroxide ions

This is a fairly obscure reaction but it is not too difficult to work out and balance. It is unusual in that hydroxide ions are acting as reducing agents.

Dark purple potassium manganate(VII) solution is slowly reduced to dark green potassium manganate(VI) solution by concentrated potassium hydroxide solution. Bubbles of oxygen gas are also given off.

The half-equation for the conversion of manganate(VII) ions to manganate(VI) ions is easy (provided, of course, that you know their formulae!).

$$MnO_4^- + e^- \longrightarrow MnO_4^{2-}$$

So what is happening to the hydroxide ions to turn them into oxygen gas?

It isn't actually very difficult to work out the half-equation directly, but suppose you want to avoid thinking and go through a standard routine:

Write down what you know, balancing the oxygens in the process:

$$2OH^- \longrightarrow O_2$$

Balance the hydrogens by adding hydrogen ions:

$$2OH^- \longrightarrow 2H^+ + O_2$$

. . . and now balance the charges:

$$2OH^- \longrightarrow 2H^+ + O_2 + 4e^-$$

Now get rid of the hydrogen ions by adding enough hydroxide ions to *both* sides of the equation:

$$2OH^- + 2OH^- \longrightarrow 2H^+ + 2OH^- + O_2 + 4e^-$$

. . . and tidy it all up!

$$4OH^- \longrightarrow 2H_2O + O_2 + 4e^-$$

Combining the half-reactions to make the ionic equation for the reaction

What we have so far is:

$$MnO_4^- + e^- \longrightarrow MnO_4^{2-}$$

$$4OH^- \longrightarrow 2H_2O + O_2 + 4e^-$$

The manganese reaction will have to happen four times in order to use up the four electrons produced by the hydroxide equation. Putting this together, you get:

$$4 \times (MnO_4^- + e^- \longrightarrow MnO_4^{2-})$$

$$1 \times (\ 4OH^- \longrightarrow 2H_2O + O_2 + 4e^-\ )$$

$$4MnO_4^- + 4OH^- \longrightarrow 4MnO_4^{2-} + 2H_2O + O_2$$

The chemistry may be unfamiliar, but working out the equation isn't too hard!

The oxidation of chromium(III) to chromium(VI)

If you add an excess of sodium hydroxide solution to a solution containing chromium(III) ions, you get a dark green solution containing the complex ion hexahydroxo-chromate(III), $Cr(OH)_6^{3-}$.

This can be oxidised to bright yellow chromate(VI) ions, CrO_4^{2-} by warming it with hydrogen peroxide solution.

We've already worked out the hydrogen peroxide half-equation where it is acting as an oxidising agent under alkaline conditions:

$$H_2O_2 + 2e^- \longrightarrow 2OH^-$$

So this time we just need to put together the half-equation for the chromium ions. What we know is:

$$Cr(OH)_6^{3-} \longrightarrow CrO_4^{2-}$$

It isn't the least bit obvious where to put hydroxide ions or water molecules, so treat it as if it were being done under acidic conditions. That way, you start by balancing the oxygens by adding water molecules.

To get 6 oxygens on each side, you need two waters on the right-hand side:

$$Cr(OH)_6^{3-} \longrightarrow CrO_4^{2-} + 2H_2O$$

Now balance the hydrogens by adding hydrogen ions:

$$Cr(OH)_6^{3-} \longrightarrow CrO_4^{2-} + 2H_2O + 2H^+$$

. . . and then balance the charges by adding electrons:

$$Cr(OH)_6^{3-} \longrightarrow CrO_4^{2-} + 2H_2O + 2H^+ + 3e^-$$

Finally, convert from acidic to alkaline conditions by adding enough hydroxide ions to both sides to turn the hydrogen ions into water:

$$Cr(OH)_6^{3-} + 2OH^- \longrightarrow CrO_4^{2-} + 2H_2O + 2H^+ + 2OH^- + 3e^-$$

. . . and tidy it all up:

$$Cr(OH)_6^{3-} + 2OH^- \longrightarrow CrO_4^{2-} + 4H_2O + 3e^-$$

Combining the half-reactions to make the ionic equation for the reaction

The two half-equations are:

$$H_2O_2 + 2e^- \longrightarrow 2OH^-$$

$$Cr(OH)_6^{3-} + 2OH^- \longrightarrow CrO_4^{2-} + 4H_2O + 3e^-$$

If you multiply one equation by 3 and the other by 2, that transfers a total of 6 electrons.

$$3 \times (H_2O_2 + 2e^- \longrightarrow 2OH^-)$$

$$2 \times (Cr(OH)_6^{3-} + 2OH^- \longrightarrow CrO_4^{2-} + 4H_2O + 3e^-)$$

$$3H_2O_2 + 2Cr(OH)_6^{3-} + 4OH^- \longrightarrow 6OH^- + 2CrO_4^{2-} + 8H_2O$$

Finally, tidy up the hydroxide ions that occur on both sides to leave the overall ionic equation:

$$3H_2O_2 + 2Cr(OH)_6^{3-} \longrightarrow 2OH^- + 2CrO_4^{2-} + 8H_2O$$

Oxidation States (Oxidation Numbers)

The oxidation states (oxidation numbers) are and how to calculate them and make use of them.

Oxidation states are straightforward to work out and to use, but it is quite difficult to define what they are in any quick way.

Explaining what oxidation states (oxidation numbers) are

Oxidation states simplify the whole process of working out what is being oxidised and what is being reduced in redox reactions. However, for the purposes of this introduction, it would be helpful if you knew about:

- Oxidation and reduction in terms of electron transfer
- Electron-half-equations

We are going to look at some examples from vanadium chemistry. If you don't know anything about vanadium, it doesn't matter in the slightest.

Vanadium forms a number of different ions - for example, V^{2+} and V^{3+}. If you think about how these might be produced from vanadium metal, the 2+ ion will be formed by oxidising the metal by removing two electrons:

$$V \longrightarrow V^{2+} + 2e^-$$

The vanadium is now said to be in an oxidation state of +2.

Removal of another electron gives the V^{3+} ion:

$$V^{2+} \longrightarrow V^{3+} + e^-$$

The vanadium now has an oxidation state of +3.

Removal of another electron gives a more unusual looking ion, VO^{2+}.

$$V^{3+} + H_2O \longrightarrow VO^{2+} + 2H^+ + e^-$$

The vanadium is now in an oxidation state of +4. Notice that the oxidation state isn't simply counting the charge on the ion (that was true for the first two cases but not for this one).

The positive oxidation state is counting the total number of electrons which have had to be removed - starting from the element.

It is also possible to remove a fifth electron to give another ion (easily confused with the one before!). The oxidation state of the vanadium is now +5.

$$VO^{2+} + H_2O \longrightarrow VO_2^{+} + 2H^{+} + e^{-}$$

Every time you oxidise the vanadium by removing another electron from it, its oxidation state increases by 1.

Fairly obviously, if you start adding electrons again the oxidation state will fall. You could eventually get back to the element vanadium which would have an oxidation state of zero.

What if you kept on adding electrons to the element? You can't actually do that with vanadium, but you can with an element like sulphur.

$$S + 2e^{-} \longrightarrow S^{2-}$$

The sulphur has an oxidation state of -2.

Oxidation state shows the total number of electrons which have been removed from an element (a positive oxidation state) or added to an element (a negative oxidation state) to get to its present state.

Oxidation involves an increase in oxidation state

Reduction involves a decrease in oxidation state

Recognising this simple pattern is the single most important thing about the concept of oxidation states. If you know how the oxidation state of an element changes during a reaction, you can instantly tell whether it is being oxidised or reduced without having to work in terms of electron-half-equations and electron transfers.

Working Out Oxidation States

You *don't* work out oxidation states by counting the numbers of electrons transferred. It would take far too long. Instead you learn some simple rules, and do some very simple sums!

- The oxidation state of an uncombined element is zero. That's obviously so, because it hasn't been either oxidised or reduced yet! This applies whatever the structure of the element - whether it is, for example, Xe or Cl_2 or S_8, or whether it has a giant structure like carbon or silicon.
- The sum of the oxidation states of all the atoms or ions in a neutral compound is zero.
- The sum of the oxidation states of all the atoms in an ion is equal to the charge on the ion.
- The more electronegative element in a substance is given a negative oxidation state. The less electronegative one is given a positive oxidation state. Remember that fluorine is the most electronegative element with oxygen second.
- Some elements almost always have the same oxidation states in their compounds:

Element	Usual oxidation state	Exceptions
Group 1 metals	always +1	
Group 2 metals	always +2	
Oxygen	usually -2	except in peroxides and F_2O
Hydrogen	usually +1	except in metal hydrides where it is -1
Fluorine	always -1	
Chlorine	usually -1	except in compounds with O or F

The reasons for the exceptions

Hydrogen in the metal hydrides

Metal hydrides include compounds like sodium hydride, NaH. In this, the hydrogen is present as a hydride ion, H^-. The oxidation state of a simple ion like hydride is equal to the charge on the ion - in this case, -1.

Alternatively, you can think of it that the sum of the oxidation states in a neutral compound is zero. Since Group 1 metals always have an oxidation state of +1 in their compounds, it follows that the hydrogen must have an oxidation state of -1 (+1 -1 = 0).

Oxygen in peroxides

Peroxides include hydrogen peroxide, H_2O_2. This is an electrically neutral compound and so the sum of the oxidation states of the hydrogen and oxygen must be zero.

Since each hydrogen has an oxidation state of +1, each oxygen must have an oxidation state of -1 to balance it.

Oxygen in F_2O

The problem here is that oxygen isn't the most electronegative element. The fluorine is more electronegative and has an oxidation state of -1. In this case, the oxygen has an oxidation state of +2.

Chlorine in compounds with fluorine or oxygen

There are so many different oxidation states that chlorine can have in these, that it is safer to simply remember that the chlorine *doesn't* have an oxidation state of -1 in them, and work out its actual oxidation state when you need it.

Don't get too bogged down in these exceptions. In most of the cases you will come across, they don't apply!

Examples of working out oxidation states

What is the oxidation state of chromium in Cr^{2+}?

That's easy! For a simple ion like this, the oxidation state is the charge on the ion - in other words: +2 (Don't forget the + sign.)

What is the oxidation state of chromium in $CrCl_3$?

This is a neutral compound so the sum of the oxidation states is zero. Chlorine has an oxidation state of -1. If the oxidation state of chromium is n:

$$n + 3(-1) = 0$$

n = +3 (Again, don't forget the + sign!)

What is the oxidation state of chromium in $Cr(H_2O)_6^{3+}$?

This is an ion and so the sum of the oxidation states is equal to the charge on the ion. There is a short-cut for working out oxidation states in complex ions like this where the metal atom is surrounded by electrically neutral molecules like water or ammonia.

The sum of the oxidation states in the attached neutral molecule must be zero. That means that you can ignore them when you do the sum. This would be essentially the same as an unattached chromium ion, Cr^{3+}. The oxidation state is +3.

What is the oxidation state of chromium in the dichromate ion, $Cr_2O_7^{2-}$?

The oxidation state of the oxygen is -2, and the sum of the oxidation states is equal to the charge on the ion. Don't forget that there are 2 chromium atoms present.

$2n + 7(-2) = -2$

$n = +6$

What is the oxidation state of copper in $CuSO_4$?

Unfortunately, it isn't always possible to work out oxidation states by a simple use of the rules above. The problem in this case is that the compound contains two elements (the copper and the sulphur) whose oxidation states can both change.

- You might recognise this as an ionic compound containing copper ions and sulphate ions, SO_4^{2-}. To make an electrically neutral compound, the copper must be present as a 2+ ion. The oxidation state is therefore +2.
- You might recognise the formula as being copper(II) sulphate. The "(II)" in the name tells you that the oxidation state is 2.
- You will know that it is +2 because you know that metals form positive ions, and the oxidation state will simply be the charge on the ion.

- *Using oxidation states*
- *In naming compounds*

You will have come across names like iron(II) sulphate and iron(III) chloride. The (II) and (III) are the oxidation states of the iron in the two compounds: +2 and +3 respectively. That tells you that they contain Fe^{2+} and Fe^{3+} ions.

This can also be extended to the negative ion. Iron(II) sulphate is $FeSO_4$. There is also a compound $FeSO_3$ with the old name of iron(II) sulphite. The modern names reflect the oxidation states of the sulphur in the two compounds.

The sulphate ion is SO_4^{2-}. The oxidation state of the sulphur is +6 (work it out!). The ion is more properly called the sulphate(VI) ion.

The sulphite ion is SO_3^{2-}. The oxidation state of the sulphur is +4 (work that out as well!). This ion is more properly called the sulphate(IV) ion. The ate ending simply shows that the sulphur is in a negative ion.

So $FeSO_4$ is properly called iron(II) sulphate(VI), and $FeSO_3$ is iron(II) sulphate(IV). In fact, because of the easy confusion between these names, the old names sulphate and sulphite are normally still used in introductory chemistry courses.

Using oxidation states to identify what's been oxidised and what's been reduced

Oxidation involves an increase in oxidation state

Reduction involves a decrease in oxidation state

In each of the following examples, we have to decide whether the reaction involves redox, and if so what has been oxidised and what reduced.

This is the reaction between magnesium and hydrochloric acid or hydrogen chloride gas:

$$Mg + 2HCl \longrightarrow MgCl_2 + H_2$$

Have the oxidation states of anything changed? Yes they have - you have two elements which are in compounds on one side of the equation and as uncombined elements on the other. Check all the oxidation states to be sure:.

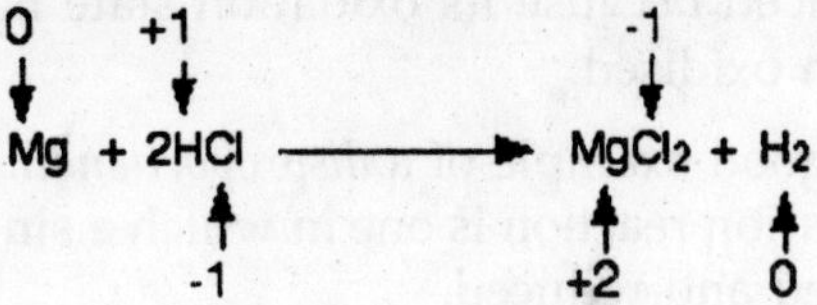

The magnesium's oxidation state has increased - it has been oxidised. The hydrogen's oxidation state has fallen - it has been reduced. The chlorine is in the same oxidation state on both sides of the equation - it hasn't been oxidised or reduced.

The reaction between sodium hydroxide and hydrochloric acid is:

$$NaOH + HCl \longrightarrow NaCl + H_2O$$

Checking all the oxidation states:

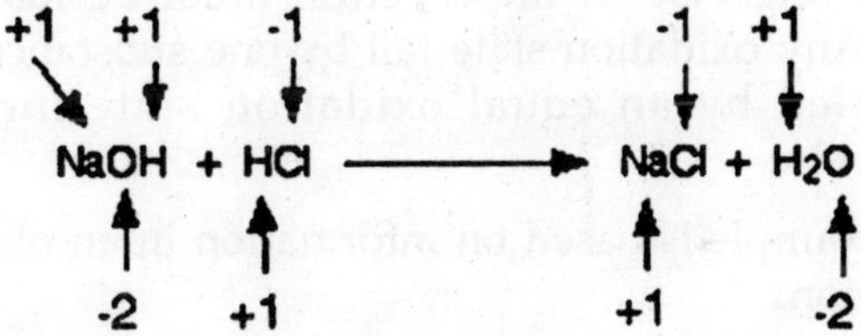

Nothing has changed. This isn't a redox reaction.

This is a sneaky one! The reaction between chlorine and cold dilute sodium hydroxide solution is:

$$2NaOH + Cl_2 \longrightarrow NaCl + NaClO + H_2O$$

Obviously the chlorine has changed oxidation state because it has ended up in compounds starting from the original element. Checking all the oxidation states shows:

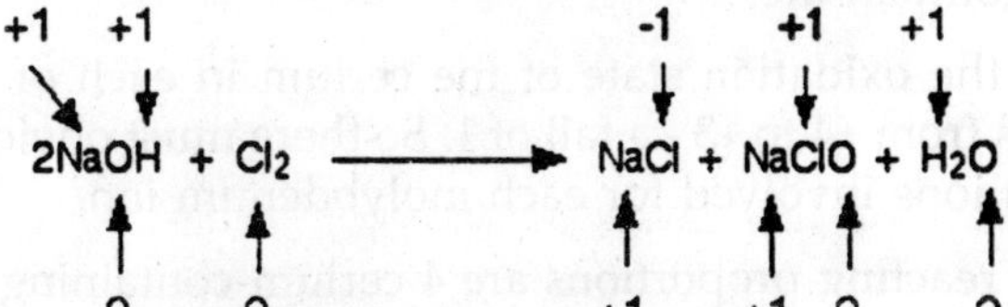

The chlorine is the *only* thing to have changed oxidation state. Has it been oxidised or reduced? Yes! Both! One atom has been reduced because its oxidation state has fallen. The other has been oxidised.

This is a good example of a *disproportionation* reaction. A disproportionation reaction is one in which a single substance is both oxidised and reduced.

Using oxidation states to work out reacting proportions

This is sometimes useful where you have to work out reacting proportions for use in titration reactions where you don't have enough information to work out the complete ionic equation.

Remember that each time an oxidation state changes by one unit, one electron has been transferred. If one substance's oxidation state in a reaction falls by 2, that means that it has gained 2 electrons.

Something else in the reaction must be losing those electrons. Any oxidation state fall by one substance must be accompanied by an equal oxidation state increase by something else.

This example is based on information in an old AQA A' level question.

Ions containing cerium in the +4 oxidation state are oxidising agents. (They are more complicated than just Ce^{4+}.) They can oxidise ions containing molybdenum from the +2 to the +6 oxidation state (from Mo^{2+} to MoO_4^{2-}). In the process the cerium is reduced to the +3 oxidation state (Ce^{3+}). What are the reacting proportions?

The oxidation state of the molybdenum is increasing by 4. That means that the oxidation state of the cerium must fall by 4 to compensate.

But the oxidation state of the cerium in each of its ions only falls from +4 to +3 - a fall of 1. So there must obviously be 4 cerium ions involved for each molybdenum ion.

The reacting proportions are 4 cerium-containing ions to 1 molybdenum ion.

Chapter 7

Acids, Bases and pH

Water is a funny substance. It makes possible much of the chemistry that goes on in our bodies and all around us. But most people take for granted the chemical properties of water. We've already learned that water molecules are constantly in motion. And keep in mind that each water molecule carries a dipole, or net charge, across the molecule.As we saw in the atomic bonding lesson, this dipole causes each molecule to behave like a little magnet with a positive and negative end. This dipole causes water molecules to be attracted to each other; the positive hydrogen is attracted to the negative oxygen of a nearby molecule. The Math Mol Water and Ice page has put together an excellent simulation of the attraction between 2 water molecules.Because the oxygen atom in water tends to monopolize the electrons in the molecule, the hydrogen protons are only loosely held to the molecule. The attraction between adjacent water molecules allows them to swap hydrogen protons. In fact, many molecules that contain hydrogen can swap protons with water molecules.When one water molecule picks up a positively charged hydrogen proton it momentarily becomes positively charged. The water molecule that looses the proton momentarily becomes negatively charged (the hydrogen's 1 electron remains behind).The result of this proton exchange is that at any given moment 2 water molecules out of every 1 billion are split into a positively charged H_3O^+ (called hydronium) ion and a negatively charge OH^- (called hydroxide) ion. In a sample of pure water, the concentration of hydronium ions is equal to 1 x 10^{-7} moles per liter (0.0000001 moles per liter). In pure water, the number of hydronium ions equals the number of hydroxide ions, so the concentration of hydroxide ions must also equal 1 x 10^{-7} moles per liter (moles per liter can be abbreviated: M). This equilibrium between hydronium and hydroxide ions can shift if we mix other substances with water. When the compound HCl is dissolved in water it separates into 2 ions: a positively charged hydrogen proton and a negatively charged chlorine ion. The positively charged hydrogen proton (H^+) combines with water and increases the concentration of H_3O^+ ions, shifting the equilibrium we discussed earlier. Some of these H_3O^+ ions recombine with the OH^- ions, and our sample becomes acidic because it contains more H_3O^+ ions than OH^- ions.The original compound that we added, HCl, is said to be an acid because it donates protons (H^+) to the mixture.On the left is a solution of pure water in which the concentration of hydronium ions equals the concentration of hydroxide ions. On the right is the same solution after addition of small amount of acid. The number of hydronium ions now exceeds the number of hydroxide ions.

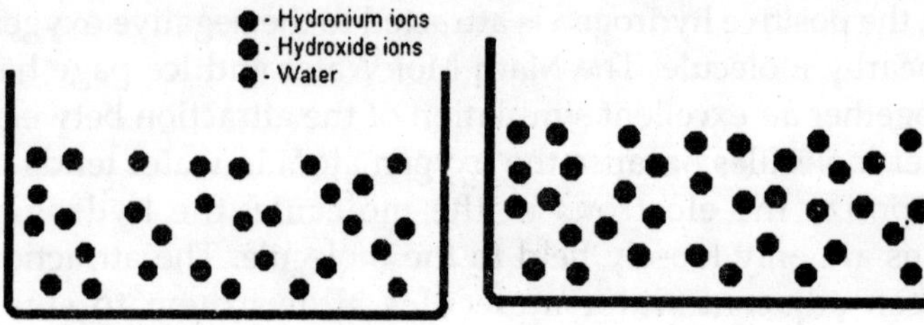

Neutral Solution: $[H_3O^+] = [OH^-]$

Acid Solution: $[H_3O^+] > [OH^-]$

An acid can be defined as a proton donor, a chemical that increases the concentration of hydronium ions in solution. Conversely, we can define a base as a proton acceptor, a chemical that reduces the concentration of hydronium ions in solution (and increases the concentration of hydroxide ions).

Acid-base chemistry is an important part of everyday life. The excess hydronium ions in acids give them interesting properties. Acids can react with metals and other materials. The strong acid HCl is produced in your stomach to help digest food. In dilute concentrations, acids are responsible for the sour taste of lemons, limes, vinegar and other substances. Bases are also very reactive. The strong base NaOH is used in many household cleaning agents such as oven cleaner and drain clog-remover. But how do we measure the concentration of an acid or base?

The acidity (or basicity) of a solution is measured using the pH scale. The pH scale corresponds to the concentration of hydronium ions in a solution. In fact, if you take the exponent of the H_3O^+ concentration and remove the negative sign, you have the pH of a solution. For example, in pure water the concentration of hydronium ions is 1×10^{-7} M. Thus, the pH of a solution of pure water is 7. The pH scale ranges from 0 to 14, where 7 is considered neutral ($[H_3O^+] = [OH^-]$), below 7 acidic and above 7 basic. The further from 7 you are on the pH scale, the more acidic or basic the solution. For example, a solution with a pH = 1 has a hydronium ion concentration of 1×10^{-1} M (0.1 M). The table below further illustrates the relationship between hydronium ion concentration and pH.

$[H_3O^+]$	pH	$[OH^-]$	Example
1×10^{0}	0	1×10^{-14}	HCl (4%)
1×10^{-1}	1	1×10^{-13}	Stomach acid
1×10^{-2}	2	1×10^{-12}	Lemon juice
1×10^{-3}	3	1×10^{-11}	Vinegar
1×10^{-4}	4	1×10^{-10}	Soda
1×10^{-5}	5	1×10^{-9}	Rainwater (unpolluted)
1×10^{-6}	6	1×10^{-8}	Milk
1×10^{-7}	7	1×10^{-7}	Pure water
1×10^{-8}	8	1×10^{-6}	Egg whites
1×10^{-9}	9	1×10^{-5}	Baking Soda
1×10^{-10}	10	1×10^{-4}	Ammonia
1×10^{-11}	11	1×10^{-3}	
1×10^{-12}	12	1×10^{-2}	Drano®
1×10^{-13}	13	1×10^{-1}	NaOH (4%)
1×10^{-14}	14	1×10^{0}	

Acids, bases, and salts

Acids are substances having the following properties:

a. Sour taste.

b. Ability to dissolve substances that water will not, e.g. zinc, iron.

c. Ability to turn blue litmus red.

d. Ability to react with alkalis, the characteristic properties of each substance disappearing ("neutralization").

Examples: Acetic acid, CH_3COOH

The 100% acid is called "glacial acetic acid" because it freezes at 17 ºC.

Hydrochloric acid: The concentrated acid is normally about 35% HCl. Anhydrous HCl is a gas, condensing to a liquid at –85 ºC.

Nitric acid : The concentrated acid is normally about 70% HNO_3.

Sulfuric acid :The concentrated acid is normally about 96% H_2SO_4.

Alkalis are substances having the following properties:

a. Solubility in water.

b. Solutions feel soapy.

c. Solutions turn red litmus blue.

d. Ability to neutralize acids.

Examples: "Soda", sodium carbonate, Na_2CO_3. "Potash", potassium carbonate, K_2CO_3."Caustic Soda", sodium hydroxide, NaOH."Caustic potash", potassium hydroxide, KOH.

Ammonia:Concentrated ammonia solution normally contains about 35% NH_3.

Soda and potash liberate carbon dioxide on treatment with acids – a further feature of the latter.

Bases are substances having the following property:

1. Ability to neutralize acids.

They thus include alkalis, but constitute a larger class, including also sparingly soluble substances that neutralize acids.

Examples: Limestone, chalk, calcium carbonate, $CaCO_3$.Quicklime, calcium oxide, CaO.Slaked lime, calcium hydroxide, $Ca(OH)_2$.

Note: quicklime is formed from limestone by heating; slaked lime is formed from quicklime by adding water;

soda and potash, once made by leaching ashes with water, can be converted into the caustic varieties by treatment with milk of lime, a suspension of slaked lime.

Salts are neutral substances (i.e. having neither the properties of an acid nor those of a base) formed by the reaction of certain acids with certain bases.

Example: Sodium chloride, NaCl, formed by the reaction of hydrochloric acid with sodium hydroxide:

$$HCl + NaOH \rightarrow NaCl + H_2O$$

Note the use of the word "certain". Sodium acetate, formed by the reaction of acetic acid with sodium hydroxide,

$$CH_3COOH + NaOH \rightarrow CH_3COONa + H_2O$$

is *not* neutral. Its solution turns red litmus blue, and it is therefore a base. When acetic acid is "neutralized" with sodium hydroxide (in the sense defined above) less sodium hydroxide is required than that needed to convert all the acetic acid into sodium acetate. Some substances can function as both an acid and a base. For example, aluminium hydroxide can react with both hydrochloric acid and sodium hydroxide:

$$Al(OH)_3 + 3HCl \rightarrow AlCl_3 + 3H_2O$$

$$Al(OH)_3 + NaOH \rightarrow Na[Al(OH)_4]$$

Substances of this type are called amphoteric

A Useful Distinction

A useful distinction can be drawn between a substance that is an acid or a base and one that *becomes* an acid or base on contact with water.Thus sulfur trioxide dissolves in water to give a solution with acidic properties. This solution is identical, however, with that obtained by dissolving sulfuric acid in water. Thus sulfur trioxide is not so much an acid but an "acid former":

$$SO_3 + H_2O \rightarrow H_2SO_4$$

Similarly quicklime is better regarded as a base-former than a base:

$$CaO + H_2O \rightarrow Ca(OH)_2$$

Acid-base character of oxides Most nonmetal oxides are acid-formers (hence Lavoisier's name for oxygen: Greek *oxus*, acid, -gen, former). Exceptions are N_2O and NO, which give neutral solutions in water.Metal oxides vary from being base-formers (e.g. Na_2O, CaO) to acid-formers (e.g. CrO_3, Mn_2O_7). This variation is determined mainly by the number of oxygen atoms (*n*) per metal atom:

low *n*	base-former
intermediate *n*	amphoteric or neutral
high *n*	acid-former

So, for example, MnO is a basic, MnO_2 is weakly amphoteric, and Mn_2O_7 is acidic. A second factor is the electronegativity of the metal (e.g. CaO is basic, ZnO is amphoteric).

Stoicheiometric Concept

This is based on a shift in the concept of "neutralization", from the complete disappearance of properties characteristic of acids and bases, to the achievement of an exact balance between the number of moles of acid and the equivalent number of moles of base. For the first of the two reactions mentioned in the first section (HCl^+ NaOH), the neutral point and the equivalence point are the same, but for the second

(CH_3COOH^+ NaOH) they are not. On the stoicheiometric conception, an *acid* may be defined as a compound of general formula H_nA which undergoes the transformation

$$H_nA + nNaOH \rightarrow Na_nA + nH_2O$$

when treated with a solution of sodium hydroxide. Here n is called the "basicity" of the acid, and A is any radical. (Note that A may contain further H's, as in the acetate radical.)

Similarly, a *base* may be defined as a compound of formula $B(OH)_m$ which undergoes the transformation

$$mHCl + B(OH)_m \rightarrow BCl_m + mH_2O$$

when treated with hydrochloric acid, m being the "acidity" of the base.

A *salt* is the product of the reaction between an acid and a base according to the appropriate stoicheiometric equation,

$$mH_nA + nB(OH)_m \rightarrow B_nA_m + mnH_2O.$$

Sodium acetate may now be called a salt, and not a base. Similarly, sodium carbonate becomes a salt – a salt of the acid carbonic acid, H_2CO_3, formed by dissolving carbon dioxide in water:

$$H_2O + CO_2 \rightleftharpoons H_2CO_3$$

$$H_2CO_3 + 2NaOH \rightarrow Na_2CO_3 + 2H_2O$$

The reaction between sodium carbonate and an acid now falls into a different class from "neutralization" (in the new sense). It now becomes a displacement reaction – the displacement of a weak acid by a strong acid from one of its salts:

$$Na_2CO_3 + 2HCl \rightarrow 2NaCl + H_2CO_3$$

(This is followed, of course by the partial decomposition of the H_2CO_3: but the displacement part of the reaction holds for a salt of any weak acid.) "Neutralization" in the new sense is a species of double decomposition. On the stoicheiometric conception of neutralization, we can now talk about the step-wise neutralization of a polybasic acid and the formation of "acid salts" and "neutral salts", e.g. $KHSO_4$ and K_2SO_4. Similarly, for a polyacidic base: e.g. $Cu(OH)_2$ forms $Cu(OH)Cl$ and $CuCl_2$, the former being a "basic" salt.The stoicheiometric conception of acids, bases, and salts is based on different types of observation from the classical conception (those that lie behind chemical formulae and equations) and sits rather loosely on the observations that lie behind the latter. Sodium carbonate, having given us the name "alkali", now becomes a salt. Care is needed to keep the two systems distinct.

Electrolytic Conception

When soluble in water, compounds that are acids, bases, or salts on the above definitions are generally good electrolytes. Thus hydrochloric acid, sodium hydroxide, and sodium chloride each behave in solution as if there are two ions per formula unit, and in electrolysis, as if each ion carries a single charge:

$HCl : H^+ + Cl^-$

$NaOH : Na^+ + OH^-$

$NaCl : Na^+ + Cl^-$

Acetic acid behaves as if it is partly dissociated into two ions according to the equilibrium

$$CH_3COOH \rightleftharpoons H^+ + CH_3COO^-$$

Sodium carbonate behaves as if it is dissociated into Na^+ and CO_3^{2-} ions, with the CO_3^{2-} ions partly reacting with the water according to the equilibrium

$$CO_3^{2-} + H_2O \rightleftharpoons HCO_3^- + OH^-$$

The latter reaction belongs to the class called "hydrolysis".These facts suggest the association of acidity with the aqueous H^+ ion and basicity with the aqueous OH^- ion. An acidic solution is then one that contains more H^+ ions than water does and a basic solution one that contains more OH^- ions. (Pure water has a very slight conductivity, attributed to low concentrations of H^+ and OH^- ions, arising from a small degree of self-ionization:

$H_2O \rightleftharpoons H^+ + OH^-$

On this basis it is possible to produce definitions of acids, bases, and salts that go some way to harmonizing the classical and stoicheiometric schemes:

- An *acid* is a compound that ionizes in water to give H^+ ions.
- A *base* is a compound that ionizes in water to give OH^- ions.
- A *salt* is a compound that, when it ionizes, does not give H^+ ions as the sole positive ions, nor OH^- ions as the sole negative ions.

On these Definitions:

a. Sodium carbonate is a salt (as on the stoicheiometric scheme), but can be also be described as a base-former (cf. the classical scheme).

b. Potassium hydrogen sulfate is both a salt (as on the stoicheiometric scheme) and an acid (as on the classical scheme), and can continue to be called an "acid salt".

c. Neutralization (classical sense) is bringing the concentration of H^+ and OH^- back to their values in pure water.

d. Neutralization in the stoicheiometric sense is unchanged.

Equations for acid-base reactions

Neutralization (stoicheiometric sense)

In the case of the reaction between a strong (i.e. completely dissociated) acid and a strong base, the equation for the reaction may be simplified as exemplified by the following:

$HCl + NaOH \rightarrow NaCl + H_2O$

$\therefore (H^+ + Cl^-) + (Na^+ + OH^-) \rightarrow (Na^+ + Cl^-) + H_2O$

$\therefore H^+ + OH^- \rightarrow H_2O$

That the reaction does not involve the counter-ions is demonstrated experimentally by the fact that the heat of neutralization of a strong acid by a strong base is the same per mole of H^+ for all such acids and bases. In the case of a reaction involving a weak acid or base, the simplified equation for the reaction is best written in terms of the unionized species. The process of simplification is then exemplified by:

$$CH_3COOH + NaOH \rightarrow CH_3COONa + H_2O$$

$$\therefore CH_3COOH + (Na^+ + OH^-) \rightarrow (CH_3COO^- + Na^+) + H_2O$$

$$\therefore CH_3COOH + OH^- \rightarrow CH_3COO^- + H_2O$$

Displacement

Displacement reactions can be simplified similarly, e.g.

$$Na_2CO_3 + 2HCl \rightarrow 2NaCl + H_2CO_3$$

$$\therefore (2Na^+ + CO_3^{2-}) + (2H^+ + 2Cl^-) \rightarrow (2Na^+ + 2Cl^-) + H_2CO_3$$

$$\therefore CO_3^{2-} + 2H^+ \rightarrow H_2CO_3$$

$$CH_3COONa + HCl \rightarrow NaCl + CH_3COOH$$

$$(CH_3COO^- + Na^+) + (H^+ + Cl^-) \rightarrow (Na^+ + Cl^-) + CH_3COOH$$

$$CH_3COO^- + H^+ \rightarrow CH_3COOH$$

In using equations like this, care needs to be taken to distinguish between those that refer to processes taking place on mixing solutions and those that refer to processes taking place within a single solution. For example, the simplified equation for the displacement of acetic acid from sodium acetate by hydrochloric acid

$$CH_3COO^- + H^+ \rightarrow CH_3COOH$$

looks, on the face of it, like the reverse of that representing the dissociation of acetic acid

$$CH_3COOH \rightleftharpoons CH_3COO^- + H^+$$

The two processes are, however, different, one taking place on mixing solutions, the other taking place within a solution:

CH_3COO^- (in solution A) + H^+ (in solution B) $\rightarrow CH_3COOH$ (in the solution obtained by mixing A & B)

CH_3COOH (in solution A) $\rightleftharpoons$ CH_3COO^- (in solution A) + H^+ (in solution A)

The equation $CH_3COO^- + H^+ \rightarrow CH_3COOH$

tends to disguise the fact that the first process involves the displacement of a weak acid by a strong acid from a salt of the weak acid, and one must be careful not to get confused.

Aqueous ammonia as a base

Aqueous ammonia is a base in the classical sense, but poses problems on the stoicheiometric scheme. On this scheme it is necessary to postulate the formation of ammonium hydroxide by the reaction

$$NH_3 + H_2O \rightarrow NH_4OH$$

the NH_4OH being the base and NH_3 the base-former. Unfortunately, no evidence for the formation of a species of formula NH_4OH has been obtained.On the electrolytic scheme, however, it is not necessary to postulate the intermediate formation of NH_4OH. The properties of aqueous ammonia can be well accounted for in terms of the equilibrium

$$NH_3 + H_2O \rightleftharpoons NH_4^+ + OH^-$$

On this scheme, NH_3 is again a base-former – it is the solution that is basic, as if there *were* a base of formula $NH_4^+OH^-$. Note, however, that tertiary alkyl amines, NR_3, on treatment with an alkyl halide, RX, give quaternary ammonium salts, $[NR_4]X$, analogous to ammonium salts, $[NH_4]X$, and these, on treatment with moist silver oxide, give the corresponding hydroxides

$$2[NR_4]X + Ag_2O + H_2O \rightarrow 2[NR_4]OH + 2AgX$$

These are very strong alkalis.

On the nature of the hydrogen ion in solution

While the hydration of all ions is acknowledged, special attention has been paid to the hydration of the hydrogen ion, for the following reason:

Anhydrous acids, unlike anhydrous bases and salts, are not ionic. For example, hydrogen chloride is a gas, and its molecules have a low polarity ($H^{0.2+}Cl^{0.2?}$). While molten sodium hydroxide and sodium chloride are good electrolytic conductors, liquid hydrogen chloride scarcely conducts electricity at all:

A few anhydrous acids are good conductors (e.g. 100% sulfuric acid), but this is unusual.Thus, in the case of the hydrogen ion, it looks as though hydration is essential for its formation. Put another way, it looks as if the H^+ ion cannot exist on its own, but only in association with one or more water molecules.The number of water molecules that it is associated with in a particular solution is difficult to determine. That one molecule is enough for its formation, however, can be demonstrated by examining the hydrates of strong acids like hydrochloric, nitric, and sulfuric. The lowest hydrates of these are as follows:

Hydrochloric acid	$HCl{\cdot}H_2O$	m.p. -15 °C
Nitric acid	$HNO_3{\cdot}H_2O$	m.p. -38 °C
Sulfuric acid	$H_2SO_4{\cdot}H_2O$	m.p. 9 °C

In the liquid state, these of course conduct electricity, and their structures in the solid state have been shown to be $H_3O^+Cl^-$, $H_3O^+NO_3^-$, and $H_3O^+HSO_4^-$ respectively.In view of the above, some chemists insist that the hydrogen ion in water should not be called the hydrogen ion, but the oxonium ion, and should not be written as H^+ but as H_3O^+. In doing this they recognize that the H_3O^+ ion will be hydrated further, but believe that no formulation less than this will do justice to the difference in properties between anhydrous and aqueous acids. In practice, however, most chemists continue to write H^+, and to call it the hydrogen ion. Whether the latter approach is lazy or the former pedantic is a matter for discussion. Some chemists further insist that compounds like HCl, HNO_3, and H_2SO_4 should not strictly be called acids. The true acids are $(H_3O)Cl$, $(H_3O)NO_3$, and $(H_3O)HSO_4$, and HCl, HNO_3, and H_2SO_4 are really acid-formers. On this basis, the latter should not be called hydrochloric acid, nitric acid, and sulfuric acid, but hydrogen chloride, hydrogen nitrate, and hydrogen sulfate. Few chemists would go as far as this, though "hydrogen chloride" has become a generally accepted name for anhydrous HCl.

IUPAC Nomenclature

For other than very common oxoacids, IUPAC prefer to name acids as compounds of H^+, e.g. hydrogen chloride. For common oxoacids, however, they recommend the traditional name, or a more systematic alternative, e.g.

	Traditional name	Alternative
H_2SO_3	sulfurous acid	trioxosulfuric acidoxosulfuric(IV) acid
H_2SO_4	sulfuric acid	tetraoxosulfuric acidoxosulfuric(VI) acid

Most users stick to the traditional names.

Lowry-Brønsted conception of acids and bases

This is a broader conception than those we have discussed so far. It is one that gets over the distinction between acids and acid-formers or bases and base-formers, though it does so at the expense of other difficulties.

On this scheme, an *acid* is a molecule or ion that supplies protons (i.e. a proton donor); a *base* is a molecule or ion that accepts protons (i.e. a proton acceptor).

This means that the molecules HCl, HNO_3, and H_2SO_4 are now acids, not acid-formers:

$$HCl + H_2O \rightarrow H_3O^+ + Cl^- \text{ etc.}$$

Similarly, the species CO_3^{2-} and NH_3 are now bases and not base-formers:

$$CO_3^{2-} + H_2O \rightleftharpoons HCO_3^- + OH^- \text{ etc.}$$

Further, the ion H_3O^+ is itself an acid, the ion OH^- is a base, and the H_2O molecule is *both* an acid *and* a base. These functions can all be seen in the self-ionization

$$2H_2O \rightleftharpoons H_3O^+ + OH^-$$

On the Lowry-Brønsted scheme, an acid-base reaction effectively involves a competition for protons representable by an equilibrium of the type

$$HX^x + Y^y \rightleftharpoons HY^{y+1} + X^{x-1}$$

where x and y represent the charges on HX and Y respectively. Here X^{x-1} is called the "conjugate base" of the acid HX^x, and HX^x the "conjugate acid" of the base X^{x-1}. For example, NH_4^+ is the conjugate acid of NH_3, and CH_3COO^- is the conjugate base of CH_3COOH, and the reaction between them involves the competition between the two bases (NH_3, CH_3COO^-) for the proton:

$$NH_4^+ + CH_3COO^- \rightleftharpoons NH_3 + CH_3COOH$$

Advantages of the Lowry-Brønsted scheme are as follows:

a. It takes in acid-formers and base-formers.

b. It is applicable to other solvents besides water. For example, in liquid ammonia, NH_4^+ acts as a proton donor, NH_2^- acts as a proton acceptor, and NH_3 functions in both ways. The Lowry-Brønsted definitions thus apply unchanged.

Disadvantages of the scheme are:

a. It loses the category of salt. The scheme is essentially dualistic in character, and has no place for a third category. Sodium chloride is regarded merely as an aggregate of ions, one of which (Na^+) is neutral, and the other (Cl^-) a base – the conjugate base of HCl.

b. The scheme also loses the idea of neutralization. Though the latter could be defined as the complete transfer of a proton from an acid to a base (cf. the stoicheiometric conception), it would include some processes for which the name neutralization hardly seems appropriate, e.g.

$$HCl + H_2O \rightarrow H_3O^+ + Cl^-$$

In view of the disadvantages, it can be questioned whether it is worth trying to extend the ideas of acid and base in the way that Lowry and Brønsted do. Chemistry might be better served if the classification of molecules and ions according to their ability to accept or donate protons – which is undoubtedly a very useful classification – were to be divorced from the classification of substances into acids and bases, and be given a terminology of its own. In fact, such terminology lies readily to hand, the terms "proton donor" and "proton acceptor" being entirely suitable for this purpose.

IUPAC are trying to dissuade chemists from calling H^+ a proton. A proton is strictly $^1H^+$. Ordinary samples of acid contain 99.985% of $^1H^+$ and 0.015% of $^2H^+$. IUPAC recommend the following names:

H hydrogen ^1H protium ^2H deuterium ^3H tritium

H^+ hydron $^1H^+$ proton $^2H^+$ deuteron $^3H^+$ triton

Classify each of the following compounds on each of the four schemes given above (classical, stoicheiometric, electrolytic, Lowry-Brønsted):

a. NaCN;

b. NH_4Cl;

c. $NaHCO_3$.

Solvent-system conception of acids and bases

The electrolytic conception of acids and bases can be extended to other ionizing solvents by means of the following definitions:

- An *acid* is a compound that ionizes to give the cation natural to the solvent.
- A *base* is a compound that ionizes to give the anion natural to the solvent.
- The ions "natural" to a solvent are those produced in its self-ionization. For some of the more important ionizing solvents, these are as follows.

Solvent	Cation	Anion
H_2O	H_3O^+	OH^-
NH_3	NH_4^+	NH_2^-
HF	H_2F^+	HF_2^-
CH_3CO_2H	$CH_3CO_2H_2^+$	$CH_3CO_2^-$
H_2SO_4	$H_3SO_4^+$	HSO_4^-
HSO_3F	$H_2SO_3F^+$	SO_3F^-
N_2O_4	NO^+	NO_3^-
BrF_3	BrF_2^+	BrF_4^-

Thus, NH_4Cl is an acid in liquid ammonia, KNH_2 is a base, $KHSO_4$ is a base in 100% sulfuric acid, etc. This extension of the ideas of acids and bases is a very useful one, and does not stretch the original conception too much. Note, however, that substances can change their character in going from one solvent to another. Thus, NH_4Cl is a salt in water, but an acid in liquid ammonia. Similarly $KHSO_4$ is an acid salt in water, and a base in 100% H_2SO_4. A third example is CH_3COOH, which is an acid-former in water

$$CH_3COOH + H_2O \rightleftharpoons H_3O^+ + CH_3COO^-$$

and a base-former in sulfuric acid

$$CH_3COOH + H_2SO_4 \rightleftharpoons H_3CO_2H_2^+ + HSO_4^-$$

This being so, it is best to specify the solvent as well as the class of substance, at least when the solvent is not water, e.g.

NH_4Cl ammonia acid

$KHSO_4$ sulfuric-acid base

Other conceptions of acids and bases

The remarks made at the end of the section on the Lowry-Brønsted definition apply with even greater force to two other definitions that have been proposed for acids and bases, namely:

	"Acid"	"Base"
Lux-Flood	Oxide ion acceptor	Oxide ion donor
Lewis	Electron-pair acceptor	Electron-pair donor

These take us a long way from the original conceptions of acid and base, and might be better divorced from these ideas completely. At the very least, they should only be used in a qualified way, as in the expressions "Lewis acid" and "Lewis base".

ACID-BASE EQUILIBRIA

The Arrhenius, Bronsted-Lowry, and Lewis theories of acids and bases, and explains the relationships between them. It also explains the concept of a conjugate pair - an acid and its conjugate base, or a base and its conjugate acid.

The Arrhenius Theory of acids and bases

a. Acids are substances which produce hydrogen ions in solution.

b. Bases are substances which produce hydroxide ions in solution.

Neutralisation happens because hydrogen ions and hydroxide ions react to produce water.

$$H^+_{(aq)} + OH^-_{(aq)} \longrightarrow H_2O_{(l)}$$

Limitations of the Theory

Hydrochloric acid is neutralised by both sodium hydroxide solution and ammonia solution. In both cases, you get a colourless solution which you can crystallise to get a white salt - either sodium chloride or ammonium chloride.

These are clearly very similar reactions. The full equations In the sodium hydroxide case, hydrogen ions from the acid are reacting with hydroxide ions from the sodium hydroxide - in line with the Arrhenius theory.However, in the ammonia case, there don't appear to be any hydroxide ions.You can get around this by saying that the ammonia reacts with the water it is dissolved in to produce ammonium ions and hydroxide ions:

$$NaOH_{(aq)} + HCl_{(aq)} \longrightarrow NaCl_{(aq)} + H_2O_{(l)}$$

$$NH_{3(aq)} + HCl_{(aq)} \longrightarrow NH_4Cl_{(aq)}$$

This is a reversible reaction, and in a typical dilute ammonia solution, about 99% of the ammonia remains as ammonia molecules. Nevertheless, there are hydroxide ions there, and we can squeeze this into the Arrhenius theory.However, this same reaction also happens between ammonia gas and hydrogen chloride gas.

$$NH_{3(aq)} + H_2O_{(l)} \rightleftharpoons NH_4^+{}_{(aq)} + OH^-{}_{(aq)}$$

In this case, there aren't any hydrogen ions or hydroxide ions in solution - because there isn't any solution. The Arrhenius theory wouldn't count this as an acid-base reaction, despite the fact that it is producing the same product as when the two substances were in solution.

The Bronsted-Lowry Theory of acids and bases

a. An acid is a proton (hydrogen ion) donor.

b. A base is a proton (hydrogen ion) acceptor.

The relationship between the Bronsted-Lowry theory and the Arrhenius theory

The Bronsted-Lowry theory doesn't go against the Arrhenius theory in any way - it just adds to it.Hydroxide ions are still bases because they accept hydrogen ions from acids and form water.An acid produces hydrogen ions in solution because it reacts with the water molecules by giving a proton to them.When hydrogen chloride gas dissolves in water to produce hydrochloric acid, the hydrogen chloride molecule gives a proton (a hydrogen ion) to a water molecule. A co-ordinate (dative covalent) bond is formed between one of the lone pairs on the oxygen and the hydrogen from the HCl. Hydroxonium ions, H_3O^+, are produced.

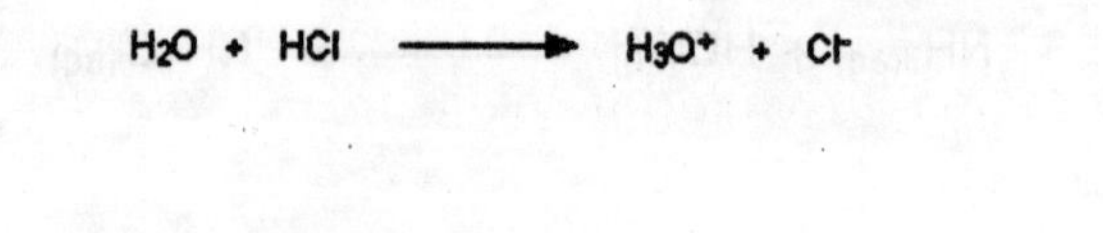

$$H_2O + HCl \longrightarrow H_3O^+ + Cl^-$$

When an acid in solution reacts with a base, what is actually functioning as the acid is the hydroxonium ion. For example, a proton is transferred from a hydroxonium ion to a hydroxide ion to make water.

$$H_3O^+_{(aq)} + OH^-_{(aq)} \longrightarrow 2H_2O_{(l)}$$

Showing the electrons, but leaving out the inner ones:

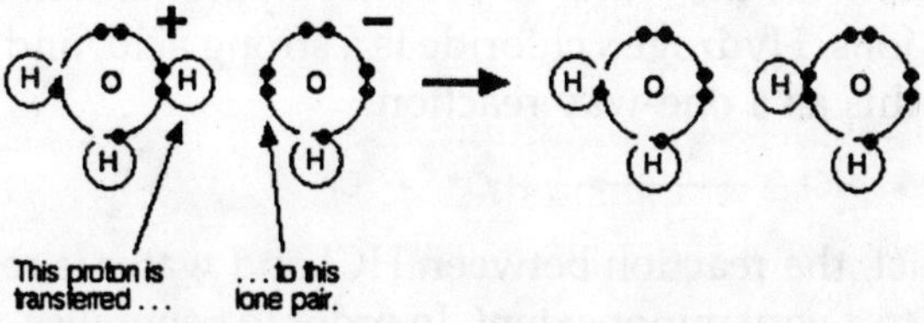

It is important to realise that whenever you talk about hydrogen ions in solution, $H^+_{(aq)}$, what you are actually talking about are hydroxonium ions.

The hydrogen chloride / ammonia problem

This is no longer a problem using the Bronsted-Lowry theory. Whether you are talking about the reaction in solution or in the gas state, ammonia is a base because it accepts a proton (a hydrogen ion). The hydrogen becomes attached to the lone pair on the nitrogen of the ammonia via a co-ordinate bond.

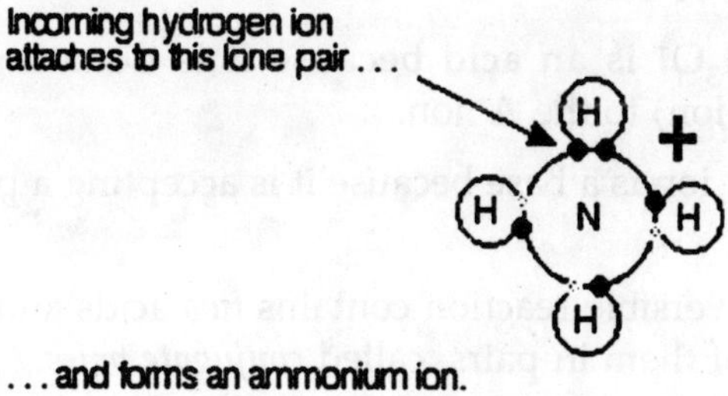

If it is in solution, the ammonia accepts a proton from a hydroxonium ion:

$$NH_{3(aq)} + H_3O^+_{(aq)} \longrightarrow NH_4{}^+_{(aq)} + H_2O_{(l)}$$

If the reaction is happening in the gas state, the ammonia accepts a proton directly from the hydrogen chloride:

$$NH_{3(g)} + HCl_{(g)} \longrightarrow NH_4^+{}_{(s)} + Cl^-{}_{(s)}$$

Either way, the ammonia acts as a base by accepting a hydrogen ion from an acid.

Conjugate Pairs

When hydrogen chloride dissolves in water, almost 100% of it reacts with the water to produce hydroxonium ions and chloride ions. Hydrogen chloride is a strong acid, and we tend to write this as a one-way reaction:

$$H_2O + HCl \longrightarrow H_3O^+ + Cl^-$$

In fact, the reaction between HCl and water is reversible, but only to a very minor extent. In order to generalise, consider an acid HA, and think of the reaction as being reversible.

$$HA + H_2O \rightleftharpoons H_3O^+ + A^-$$

Thinking about the *forward reaction:*

The HA is an acid because it is donating a proton (hydrogen ion) to the water.

The water is a base because it is accepting a proton from the HA.

But there is also a *back reaction* between the hydroxonium ion and the A^- ion:

The H_3O^+ is an acid because it is donating a proton (hydrogen ion) to the A^- ion.

The A^- ion is a base because it is accepting a proton from the H_3O^+.

The reversible reaction contains *two* acids and *two* bases. We think of them in pairs, called *conjugate pairs.*

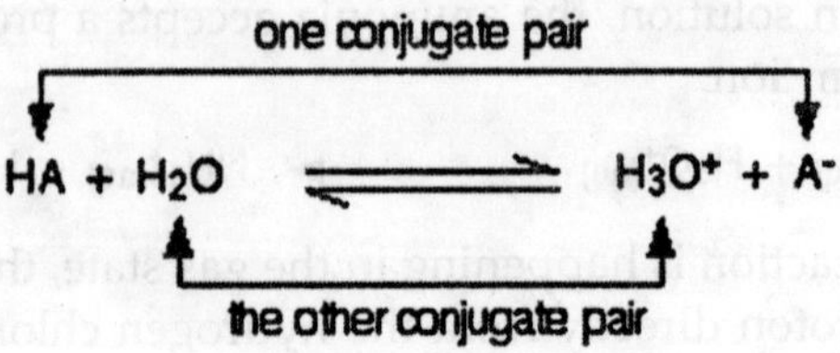

When the acid, HA, loses a proton it forms a base, A^-. When the base, A^-, accepts a proton back again, it obviously refoms the acid, HA. These two are a conjugate pair.

Members of a conjugate pair differ from each other by the presence or absence of the transferable hydrogen ion.

If you are thinking about HA as the acid, then A^- is its conjugate base.If you are thinking about A^- as the base, then HA is its conjugate acid.

The water and the hydroxonium ion are also a conjugate pair. Thinking of the water as a base, the hydroxonium ion is its conjugate acid because it has the extra hydrogen ion which it can give away again.Thinking about the hydroxonium ion as an acid, then water is its conjugate base. The water can accept a hydrogen ion back again to reform the hydroxonium ion.

A second example of conjugate pairs

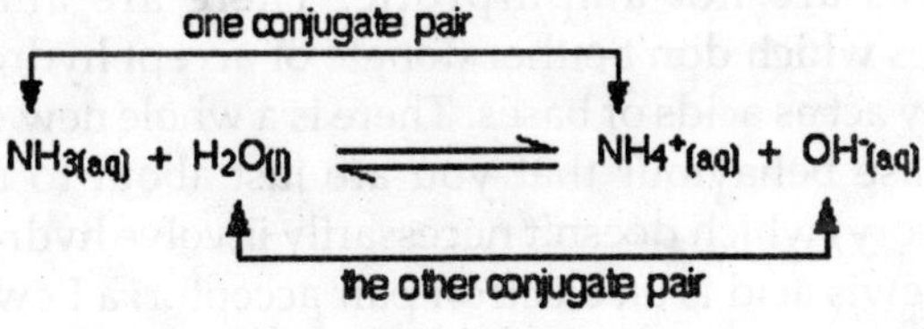

This is the reaction between ammonia and water that we looked at earlier:

Think first about the forward reaction. Ammonia is a base because it is accepting hydrogen ions from the water. The ammonium ion is its conjugate acid - it can release that hydrogen ion again to reform the ammonia.The water is acting as an acid, and its conjugate base is the hydroxide ion. The hydroxide ion can accept a hydrogen ion to reform the water.Looking at it from the other side, the ammonium ion is an acid, and ammonia is its conjugate base. The hydroxide ion is a base and water is its conjugate acid.

Amphoteric Substances

You may possibly have noticed (although probably not!) that in one of the last two examples, water was acting as a base, whereas in the other one it was acting as an acid.A substance which can act as either an acid or a base is described as being *amphoteric*.

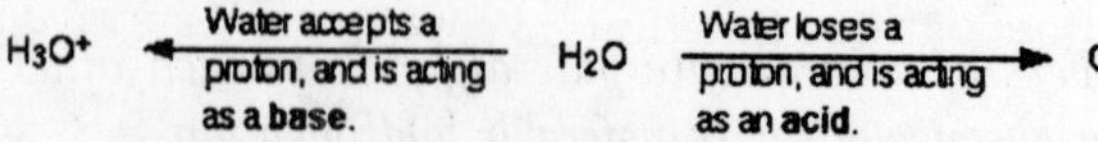

An *amphiprotic* substance is one which can both donate hydrogen ions (protons) and also accept them. Water is a good example of such a compound. The water acts as both an acid (donating hydrogen ions) and as a base (by accepting them). The "protic" part of the word refers to the hydrogen ions (protons) either being donated or accepted. Other examples of amphiprotic compounds are amino acids, and ions like HSO_4^- (which can lose a hydrogen ion to form sulphate ions or accept one to form sulphuric acid).But as well as being amphiprotic, these compounds are also *amphoteric*. Amphoteric means that they have reactions as both acids and bases. All amphiprotic substances are also amphoteric but all amphoteric substances are not amphiprotic. There are amphoteric substances which don't either donate or accept hydrogen ions when they act as acids or bases. There is a whole new definition of acid-base behaviour that you are just about to meet (the Lewis theory) which doesn't necessarily involve hydrogen ions at all.A Lewis acid is an electron pair acceptor; a Lewis base is an electron pair donor Some metal oxides (like aluminium oxide) are amphoteric - they react both as acids and bases. For example, they react as bases because the oxide ions accept hydrogen ions to make water. That's not a problem as far as the definition of amphiprotic is concerned - but the reaction as an acid is. The aluminium oxide doesn't contain any hydrogen ions to donate! But aluminium oxide reacts with bases like sodium hydroxide solution to form complex aluminate ions.You can think of lone pairs on hydroxide ions as forming dative covalent (coordinate) bonds with empty orbitals in the aluminium ions. The aluminium ions are accepting lone pairs (acting as a Lewis acid). So aluminium oxide can act as both an acid and a base - and so is amphoteric. But it *isn't* amphiprotic because *both* of the acid reaction and the base reaction don't involve hydrogen ions.Without once using the term amphiprotic. The term amphoteric takes in all the cases of substances functioning as both acids and bases without exception. The term amphiprotic can only be used where both of these functions involve transference of hydrogen ions - in other words, it can only be used if you are limited to talking about the Bronsted-Lowry theory.

The Lewis Theory of Acids and Bases

An acid is an electron pair acceptor.

A base is an electron pair donor.

The relationship between the Lewis theory and the Bronsted-Lowry theory

It is easiest to see the relationship by looking at exactly what Bronsted-Lowry bases do when they accept hydrogen ions. Three Bronsted-Lowry bases we've looked at are hydroxide ions, ammonia and water, and they are typical of all the rest.

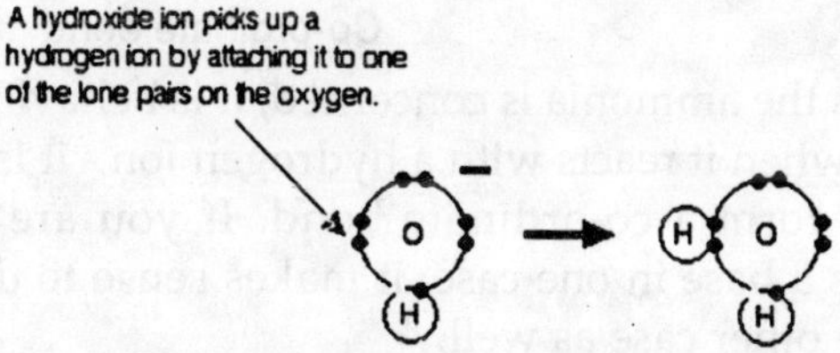

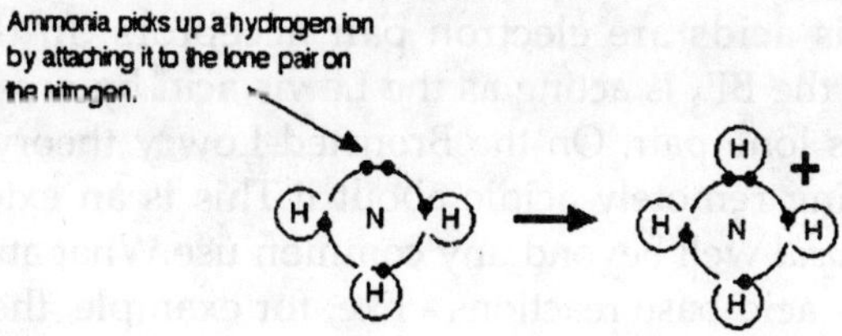

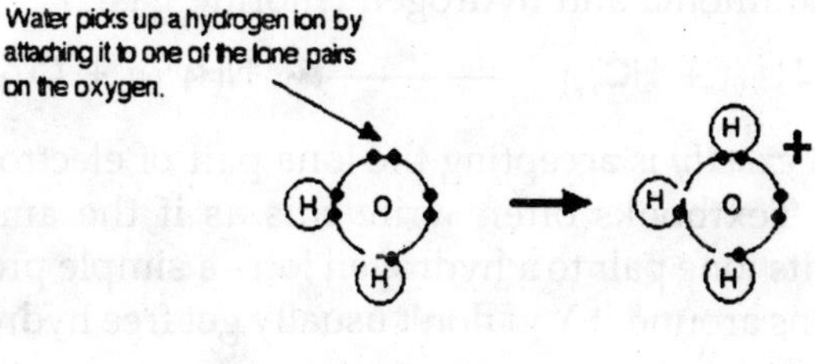

The Bronsted-Lowry theory says that they are acting as bases because they are combining with hydrogen ions. The reason they are combining with hydrogen ions is that they have lone pairs of electrons - which is what the Lewis theory says. The two are entirely consistent.On the Lewis theory, any reaction in which the ammonia or water used their lone pairs of electrons to form a co-ordinate bond would be counted as them acting as a base.Here is a reaction which you will find talked about on the page dealing with co-ordinate bonding. Ammonia reacts with BF_3 by using its lone pair to form a co-ordinate bond with the empty orbital on the boron.

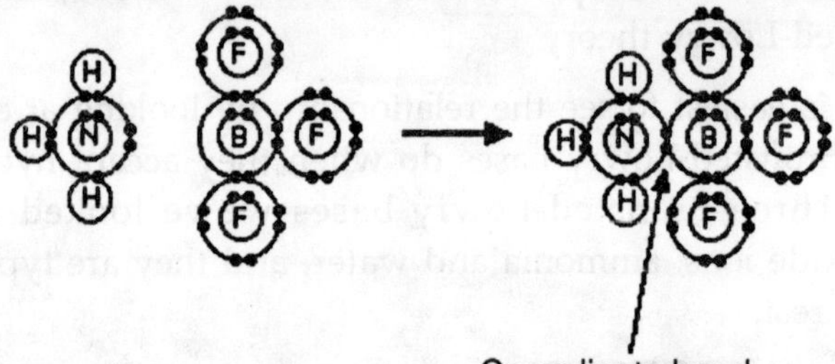

As far as the ammonia is concerned, it is behaving exactly the same as when it reacts with a hydrogen ion - it is using its lone pair to form a co-ordinate bond. If you are going to describe it as a base in one case, it makes sense to describe it as one in the other case as well.

Lewis acids are electron pair acceptors. In the above example, the BF_3 is acting as the Lewis acid by accepting the nitrogen's lone pair. On the Bronsted-Lowry theory, the BF_3 has nothing remotely acidic about it.This is an extension of the term *acid* well beyond any common use.What about more obviously acid-base reactions - like, for example, the reaction between ammonia and hydrogen chloride gas?

$$NH_{3(g)} + HCl_{(g)} \longrightarrow NH_4^+{}_{(s)} + Cl^-{}_{(s)}$$

What *exactly* is accepting the lone pair of electrons on the nitrogen. Textbooks often write this as if the ammonia is donating its lone pair to a hydrogen ion - a simple proton with no electrons around it.You don't usually get free hydrogen ions in chemical systems. They are so reactive that they are always attached to something else. There aren't any uncombined hydrogen ions in HCl.

There isn't an empty orbital anywhere on the HCl which can accept a pair of electrons.

Why, then, is the HCl a Lewis acid?

Chlorine is more electronegative than hydrogen, and that means that the hydrogen chloride will be a polar molecule. The electrons in the hydrogen-chlorine bond will be attracted towards the chlorine end, leaving the hydrogen slightly positive and the chlorine slightly negative.

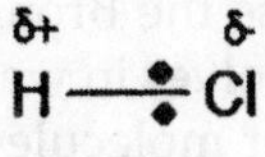

The lone pair on the nitrogen of an ammonia molecule is attracted to the slightly positive hydrogen atom in the HCl. As it approaches it, the electrons in the hydrogen-chlorine bond are repelled still further towards the chlorine.A co-ordinate bond is formed between the nitrogen and the hydrogen, and the chlorine breaks away as a chloride ion.This is best shown using the "curly arrow" notation commonly used in organic reaction mechanisms.

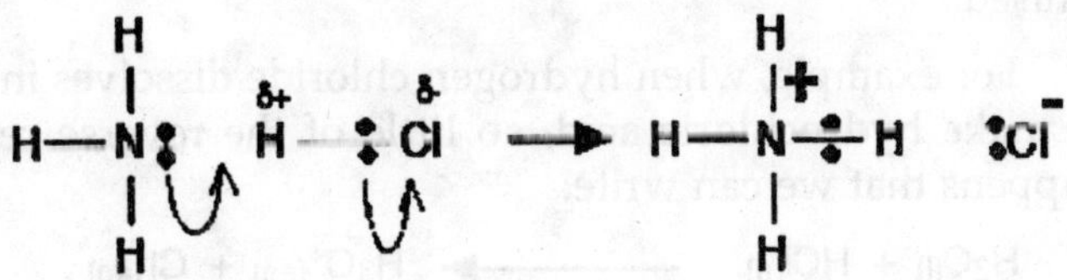

The whole HCl molecule is acting as a Lewis acid. It is accepting a pair of electrons from the ammonia, and in the process it breaks up. Lewis acids don't necessarily have to have an existing empty orbital.

- A Lewis acid is an electron pair acceptor.
- A Lewis base is an electron pair donor.

For all general purposes, stick with the Bronsted-Lowry theory.

STRONG AND WEAK ACIDS

The terms strong and weak as applied to acids. As a part of this it defines and explains what is meant by pH, K_a and pK_a.It is important that you don't confuse the words *strong* and *weak* with the terms *concentrated* and *dilute*.As you will see below, the strength of an acid is related to the proportion of it which has reacted with water to produce ions. The concentration tells you about how much of the original acid is dissolved in the solution.It is perfectly possible to have a concentrated solution of a weak acid, or a dilute solution of a strong acid.

Strong Acids

We are going to use the Bronsted-Lowry definition of an acid. When an acid dissolves in water, a proton (hydrogen ion) is transferred to a water molecule to produce a hydroxonium ion and a negative ion depending on what acid you are starting from.

In the general case . . .

$$HA + H_2O \rightleftharpoons H_3O^+ + A^-$$

These reactions are all reversible, but in some cases, the acid is so good at giving away hydrogen ions that we can think of the reaction as being one-way. The acid is virtually 100% ionised.

For example, when hydrogen chloride dissolves in water to make hydrochloric acid, so little of the reverse reaction happens that we can write:

$$H_2O_{(l)} + HCl_{(g)} \longrightarrow H_3O^+_{(aq)} + Cl^-_{(aq)}$$

At any one time, virtually 100% of the hydrogen chloride will have reacted to produce hydroxonium ions and chloride ions. Hydrogen chloride is described as a *strong acid.A strong acid is one which is virtually 100% ionised in solution.*

Other common strong acids include sulphuric acid and nitric acid.You may find the equation for the ionisation written in a simplified form:

$$HCl_{(aq)} \longrightarrow H^+_{(aq)} + Cl^-_{(aq)}$$

This shows the hydrogen chloride dissolved in the water splitting to give hydrogen ions in solution and chloride ions in solution. This version is often used in this work just to make things look easier. If you use it, remember that the water is actually involved, and that when you write $H^+_{(aq)}$ what you really mean is a hydroxonium ion, H_3O^+.

Strong acids and pH

pH is a measure of the concentration of hydrogen ions in a solution. Strong acids like hydrochloric acid at the sort of concentrations you normally use in the lab have a pH around 0 to 1. The lower the pH, the higher the concentration of hydrogen ions in the solution.

Defining pH

Notice the negative sign. The square brackets mean "concentration in mol dm^{-3}."

$$pH = -\log_{10}[H^+]$$

Read this as "log to base 10".

Suppose you had to work out the pH of 0.1 mol dm^{-3} hydrochloric acid. All you have to do is work out the concentration of the hydrogen ions in the solution, and then use your calculator to convert it to a pH.

Hydrochloric acid is a strong acid - virtually 100% ionised. Each mole of HCl reacts with the water to give 1 mole of hydrogen ions and 1 mole of chloride ionsThat means that if the concentration of the acid is 0.1 mol dm^{-3}, then the concentration of hydrogen ions is also 0.1 mol dm^{-3}.Use your calculator to convert this into pH. My calculator wants me to enter 0.1, and then press the "log" button. Yours might want you to do it in a different order. You need to find out!

$\log_{10}[0.1] = -1$

But $pH = -\log_{10}[0.1]$

- (-1) = 1

The pH of this acid is 1.

Weak acids

A weak acid is one which doesn't ionise fully when it is dissolved in water.Ethanoic acid is a typical weak acid. It reacts with water to produce hydroxonium ions and ethanoate ions, but the back reaction is more successful than the forward one. The ions react very easily to reform the acid and the water.

$$CH_3COOH + H_2O \rightleftharpoons CH_3COO^- + H_3O^+$$

At any one time, only about 1% of the ethanoic acid molecules have converted into ions. The rest remain as simple ethanoic acid molecules.Most organic acids are weak. Hydrogen fluoride (dissolving in water to produce hydrofluoric acid) is a weak inorganic acid that you may come across elsewhere.

Comparing the strengths of weak acids

The position of equilibrium of the reaction between the acid and water varies from one weak acid to another. The further to the left it lies, the weaker the acid is.

$$HA + H_2O \rightleftharpoons H_3O^+ + A^-$$

The acid dissociation constant, K_a

You can get a measure of the position of an equilibrium by writing an equilibrium constant for the reaction. The lower the value for the constant, the more the equilibrium lies to the left.The dissociation (ionisation) of an acid is an example of a homogeneous reaction. Everything is present in the same phase - in this case, in solution in water. You can therefore write a simple expression for the equilibrium constant, K_c.

Here is the equilibrium again:

$$HA + H_2O \rightleftharpoons H_3O^+ + A^-$$

You might *expect* the equilibrium constant to be written as:

$$K_c = \frac{[H_3O^+]\,[A^-]}{[HA]\,[H_2O]}$$

However, if you think about this carefully, there is something odd about it.At the bottom of the expression, you have a term for the concentration of the water in the solution. That's not a problem - except that the number is going to be very large compared with all the other numbers.In 1 dm^3 of solution, there are going to be about 55 moles of water.If you had a weak acid with a concentration of about 1 mol dm^{-3}, and only about 1% of it reacted with the water, the number of moles of water is only going to fall by about 0.01. In other words, if the acid is weak the concentration of the water is virtually constant.In that case, there isn't a lot of point in including it in the expression as if it were a variable. Instead, a new equilibrium constant is defined which leaves it out. This new equilibrium constant is called K_a.

$$K_a = \frac{[H_3O^+]\,[A^-]}{[HA]}$$

You may find the K_a expression written differently if you work from the simplified version of the equilibrium reaction:

$$HA_{(aq)} \rightleftharpoons H^+_{(aq)} + A^-_{(aq)}$$

$$K_a = \frac{[H^+]\,[A^-]}{[HA]}$$

This may be written with or without state symbols.

It is actually exactly the same as the previous expression for K_a! Remember that although we often write H^+ for hydrogen ions in solution, what we are actually talking about are hydroxonium ions.This second version of the K_a expression isn't as precise as the first one, but your examiners may well accept it.To take a specific common example, the equilibrium for the dissociation of ethanoic acid is properly written as:

$$CH_3COOH + H_2O \rightleftharpoons CH_3COO^- + H_3O^+$$

The K_a expression is:

$$K_a = \frac{[CH_3COO^-][H_3O^+]}{[CH_3COOH]}$$

If you are using the simpler version of the equilibrium . .

$$CH_3COOH \rightleftharpoons CH_3COO^- + H^+$$

. . . the K_a expression is:

$$K_a = \frac{[CH_3COO^-][H^+]}{[CH_3COOH]}$$

The table shows some values of K_a for some simple acids:

Acid	K_a (mol dm^{-3})
hydrofluoric acid	5.6×10^{-4}
methanoic acid	1.6×10^{-4}
ethanoic acid	1.7×10^{-5}
hydrogen sulphide	8.9×10^{-8}

These are all weak acids because the values for K_a are very small. They are listed in order of decreasing acid strength - the K_a values get smaller as you go down the table.To avoid this, the numbers are often converted into a new, easier form, called pK_a.

pK_a

pK_a bears exactly the same relationship to K_a as pH does to the hydrogen ion concentration:

$$pK_a = -\log_{10} K_a$$

If you use your calculator on all the K_a values in the table above and convert them into pK_a values, you get:

Acid	**K_a (mol dm^{-3})**	**pK_a**
hydrofluoric acid	5.6×10^{-4}	3.3
methanoic acid	1.6×10^{-4}	3.8
ethanoic acid	1.7×10^{-5}	4.8
Hydrogen sulphide	8.9×10^{-8}	7.1

pK_a doesn't have any units.

Notice that the weaker the acid, the larger the value of pK_a. It is now easy to see the trend towards weaker acids as you go down the table.

The lower the value for pK_a, the stronger the acid.

The higher the value for pK_a, the weaker the acid.

THE IONIC PRODUCT FOR WATER, K_w

The ionic product for water it looks at how the ionic product varies with temperature, and how that determines the pH of pure water at different temperatures.

K_w and pK_w

Water molecules can function as both acids and bases. One water molecule (acting as a base) can accept a hydrogen ion from a second one (acting as an acid). This will be happening anywhere there is even a trace of water - it doesn't have to be pure.A hydroxonium ion and a hydroxide ion are formed.

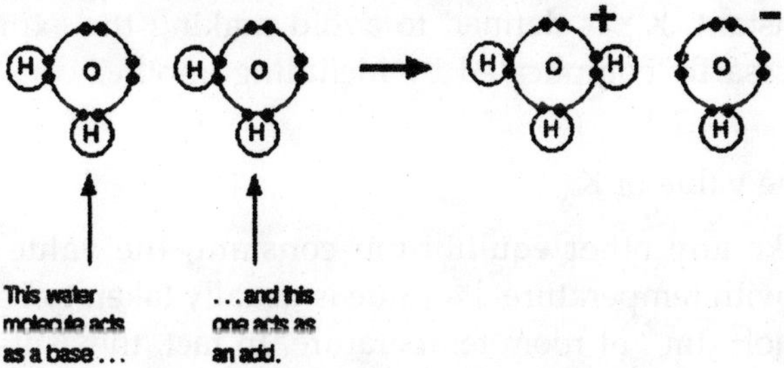

However, the hydroxonium ion is a very strong acid, and the hydroxide ion is a very strong base. As fast as they are formed, they react to poduce water again.The net effect is that an equilibrium is set up.

$$2H_2O_{(l)} \rightleftharpoons H_3O^+_{(aq)} + OH^-_{(aq)}$$

At any one time, there are incredibly small numbers of hydroxonium ions and hydroxide ions present. we shall calculate the concentration of hydroxonium ions present in pure water. It turns out to be 1.00×10^{-7} mol dm^{-3} at room temperature.You may well find this equilibrium written in a simplified form:

$$H_2O_{(l)} \rightleftharpoons H^+_{(aq)} + OH^-_{(aq)}$$

This is OK provided you remember that $H^+_{(aq)}$ actually refers to a hydroxonium ion.

Defining the ionic product for water, K_w

K_w is essentially just an equilibrium constant for the reactions shown. You may meet it in two forms:

Based on the fully written equilibrium . . .

$$K_w = [H_3O^+][OH^-]$$

. . . or on the simplified equilibrium:

$$K_w = [H^+][OH^-]$$

The water isn't written on the bottom of these equilibrium constant expressions. So little of the water is ionised at any one time, that its concentration remains virtually unchanged - a constant. K_w is defined to avoid making the expression unnecessarily complicated by including another constant in it.

The value of K_w

Like any other equilibrium constant, the value of K_w varies with temperature. Its value is usually taken to be 1.00 x 10^{-14} mol^2 dm^{-6} at room temperature. In fact, this is its value at a bit less than 25°C.

The units of K_w: K_w is found by multiplying two concentration terms together. Each of these has the units of mol dm^{-3}.Multiplying mol dm^{-3} x mol dm^{-3} gives you the units above.

pK_w

The relationship between K_w and pK_w is exactly the same as that between K_a and pK_a, or $[H^+]$ and pH.

$$pK_w = - \log_{10} K_w$$

The K_w value of 1.00×10^{-14} $mol^2\ dm^{-6}$ at room temperature gives you a pK_w value of 14. Try it on your calculator! Notice that pK_w doesn't have any units.

The pH of pure water

That question is actually misleading! In fact, pure water only has a pH of 7 at a particular temperature - the temperature at which the K_w value is 1.00×10^{-14} $mol^2\ dm^{-6}$.

To find the pH you need first to find the hydrogen ion concentration. In pure water at room temperature the K_w value tells you that:

$[H^+]\,[OH^-] = 1.00 \times 10^{-14}$

But in pure water, the hydrogen ion (hydroxonium ion) concentration must be equal to the hydroxide ion concentration. For every hydrogen ion formed, there is a hydroxide ion formed as well.

That means that you can replace the $[OH^-]$ term in the K_w expression by another $[H^+]$.

$[H^+]^2 = 1.00 \times 10^{-14}$

Taking the square root of each side gives:

$[H^+] = 1.00 \times 10^{-7}$ mol dm^{-3}

Converting that into pH:

$pH = -\log_{10}[H^+]$

$pH = 7$

That's where the familiar value of 7 comes from.

The variation of the pH of pure water with temperature

The formation of hydrogen ions (hydroxonium ions) and hydroxide ions from water is an endothermic process. Using the simpler version of the equilibrium:

$$H_2O_{(l)} \rightleftharpoons H^+_{(aq)} + OH^-_{(aq)}$$

The forward reaction absorbs heat.

According to Le Chatelier's Principle, if you make a change to the conditions of a reaction in dynamic equilibrium, the position of equilibrium moves to counter the change you have made.

According to Le Chatelier, if you increase the temperature of the water, the equilibrium will move to lower the temperature again. It will do that by absorbing the extra heat.

That means that the forward reaction will be favoured, and more hydrogen ions and hydroxide ions will be formed. The effect of that is to increase the value of K_w as temperature increases.

The table below shows the effect of temperature on K_w. For each value of K_w, a new pH has been calculated using the same method as above. It might be useful if you were to check these pH values yourself.

T (°C)	K_w (mol² dm⁻⁶)	pH
0	0.114×10^{-14}	7.47
10	0.293×10^{-14}	7.27
20	0.681×10^{-14}	7.08
25	1.008×10^{-14}	7.00
30	1.471×10^{-14}	6.92
40	2.916×10^{-14}	6.77
50	5.476×10^{-14}	6.63
100	51.3×10^{-14}	6.14

The pH of pure water falls as the temperature increases.The pH falls as temperature increases, doesnot mean that water becomes more acidic at higher temperatures.

A solution is acidic if there is an excess of hydrogen ions over hydroxide ions. In the case of pure water, there are always the same number of hydrogen ions and hydroxide ions. That means that the water remains neutral - even if its pH changes.The problem is that we are all so familiar with 7 being the pH of pure water, that anything else feels really strange. Remember that you calculate the neutral value of pH from K_w. If that changes, then the neutral value for pH changes as well.At 100°C, the pH of pure water is 6.14. That is the neutral point on the pH scale at this higher temperature. A solution with a pH of 7 at this temperature is slightly alkaline because its pH is a bit higher than the neutral value of 6.14.Similarly, you can argue that a solution with a pH of 7 at 0°C is slightly acidic, because its pH is a bit lower than the neutral value of 7.47 at this temperature.

STRONG AND WEAK BASES

The terms strong and weak as applied to bases. As a part of this it defines and explains K_b and pK_b.We are going to use the Bronsted-Lowry definition of a base as a substance which accepts hydrogen ions (protons).The usual way of comparing the strengths of bases is to see how readily they produce hydroxide ions in solution. This may be because they already contain hydroxide ions, or because they take hydrogen ions from water molecules to produce hydroxide ions.

Strong Bases

A strong base is something like sodium hydroxide or potassium hydroxide which is fully ionic. You can think of the compound as being 100% split up into metal ions and hydroxide ions in solution.Each mole of sodium hydroxide dissolves to give a mole of hydroxide ions in solution.

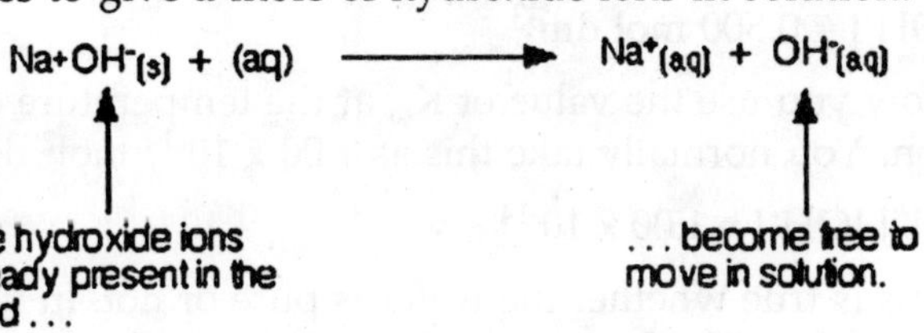

Some strong bases like calcium hydroxide aren't very soluble in water. That doesn't matter - what does dissolve is still 100% ionised into calcium ions and hydroxide ions. Calcium hydroxide still counts as a strong base because of that 100% ionisation.

$$pH = -\log_{10}[H^+]$$

Since pH is a measure of hydrogen ion concentration, how can a solution which contains hydroxide ions have a pH? To understand this, you need to know about the ionic product for water.

Wherever there is water, an equilibrium is set up. Using the simplified version of this equilibrium:

$$H_2O_{(l)} \rightleftharpoons H^+_{(aq)} + OH^-_{(aq)}$$

In the presence of extra hydroxide ions from, say, sodium hydroxide, the equilibrium is still there, but the position of equilibrium has been shifted well to the left according to Le Chatelier's Principle.

There will be far fewer hydrogen ions than there are in pure water, but there will still be hydrogen ions present. The pH is a measure of the concentration of these.

Work out the concentration of the hydroxide ions.

Use K_w to work out the hydrogen ion concentration.

Convert the hydrogen ion concentration to a pH.

To find the pH of 0.500 mol dm^{-3} sodium hydroxide solution:

Because the sodium hydroxide is fully ionic, each mole of it gives that same number of moles of hydroxide ions in solution.

$[OH^-] = 0.500$ mol dm^{-3}

Now you use the value of K_w at the temperature of your solution. You normally take this as 1.00×10^{-14} mol^2 dm^{-6}.

$[H^+] [OH^-] = 1.00 \times 10^{-14}$

This is true whether the water is pure or not. In this case we have a value for the hydroxide ion concentration. Substituting that gives:

$[H^+] \times 0.500 = 1.00 \times 10^{-14}$

If you solve that for $[H^+]$, and then convert it into pH, you get a pH of 13.7.

Weak Bases

Ammonia is a typical weak base. Ammonia itself obviously doesn't contain hydroxide ions, but it reacts with water to produce ammonium ions and hydroxide ions.

$$NH_{3(aq)} + H_2O_{(l)} \rightleftharpoons NH_4{}^+{}_{(aq)} + OH^-{}_{(aq)}$$

However, the reaction is reversible, and at any one time about 99% of the ammonia is still present as ammonia molecules. Only about 1% has actually produced hydroxide ions.A weak base is one which doesn't convert fully into hydroxide ions in solution. When a weak base reacts with water, the position of equilibrium varies from base to base. The further to the left it is, the weaker the base.

$$B:_{(aq)} + H_2O_{(l)} \rightleftharpoons B:H^+{}_{(aq)} + OH^-{}_{(aq)}$$

Shows the lone pair on the base which accepts the hydrogen ion

You can get a measure of the position of an equilibrium by writing an equilibrium constant for the reaction. The lower the value for the constant, the more the equilibrium lies to the left.In this case the equilibrium constant is called K_b. This is defined as:

$$K_b = \frac{[B:H^+]\,[OH^-]}{[B:]}$$

pK_b

The relationship between K_b and pK_b is exactly the same as all the other "p" terms in this topic:

$$pK_b = -\log_{10} K_b$$

The table shows some values for K_b and pK_b for some weak bases.

base	K_b (mol dm^{-3})	pK_b
$C_6H_5NH_2$	4.17×10^{-10}	9.38
NH_3	1.78×10^{-5}	4.75
CH_3NH_2	4.37×10^{-4}	3.36
$CH_3CH_2NH_2$	5.37×10^{-4}	3.27

As you go down the table, the value of K_b is increasing. That means that the bases are getting stronger.As K_b gets bigger, pK_b gets smaller. The lower the value of pK_b, the stronger the base.This is exactly in line with the corresponding term for acids, pK_a - the smaller the value, the stronger the acid.

PH (TITRATION) CURVES

When you carry out a simple acid-base titration, you use an indicator to tell you when you have the acid and alkali mixed in exactly the right proportions to "neutralise" each other. When the indicator changes colour, this is often described as the *end point* of the titration.In an ideal world, the colour change would happen when you mix the two solutions together in exactly equation proportions. That particular mixture is known as the *equivalence point*.For example, if you were titrating sodium hydroxide solution with hydrochloric acid, both with a concentration of 1 mol dm^{-3}, 25 cm^3 of sodium hydroxide solution would need exactly the same volume of the acid - because they react 1 : 1 according to the equation.

$$NaOH_{(aq)} + HCl_{(aq)} \longrightarrow NaCl_{(aq)} + H_2O_{(l)}$$

In this particular instance, this would also be the *neutral point* of the titration, because sodium chloride solution has a pH of 7. But that isn't necessarily true of all the salts you might get formed.For example, if you titrate ammonia solution with hydrochloric acid, you would get ammonium chloride formed. The ammonium ion is slightly acidic, and so pure ammonium chloride has a slightly acidic pH.That means that at the equivalence point (where you had mixed the solutions in the correct proportions according to the equation), the solution wouldn't actually be neutral. To use the term "neutral point" in this context would be misleading.Similarly, if you titrate sodium hydroxide solution with ethanoic acid, at the equivalence point the pure sodium ethanoate formed has a slightly alkaline pH because the ethanoate ion is slightly basic.

The term "neutral point" is best avoided.

The term "equivalence point" means that the solutions have been mixed in exactly the right proportions according to the equation.

The term "end point" is where the indicator changes colour. As you will see on the page about indicators, that isn't necessarily exactly the same as the equivalence point.

Simple pH curves

All the following titration curves are based on both acid and alkali having a concentration of 1 mol dm^{-3}. In each case, you start with 25 cm^3 of one of the solutions in the flask, and the other one in a burette.Although you normally run the acid from a burette into the alkali in a flask, you may need to know about the titration curve for adding it the other way around as well. Alternative versions of the curves have been described in most cases.

Titration Curves

We'll take hydrochloric acid and sodium hydroxide as typical of a strong acid and a strong base.

$$NaOH_{(aq)} + HCl_{(aq)} \longrightarrow NaCl_{(aq)} + H_2O_{(l)}$$

Running acid into the alkali

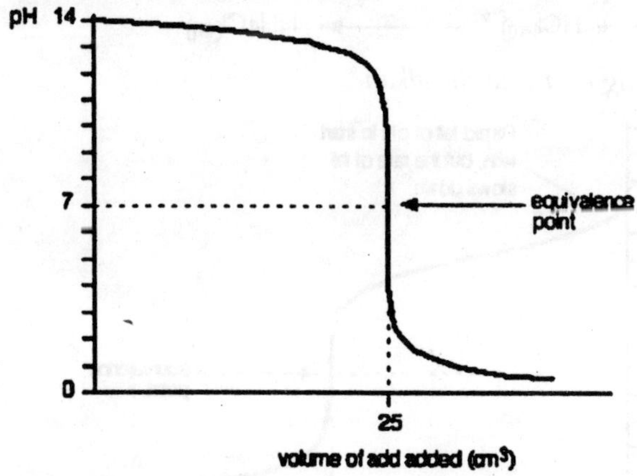

The pH only falls a very small amount until quite near the equivalence point. Then there is a really steep plunge. If you calculate the values, the pH falls all the way from 11.3 when you have added 24.9 cm^3 to 2.7 when you have added 25.1 cm^3.

Running Alkali into the Acid

This is very similar to the previous curve except, of course, that the pH starts off low and increases as you add more sodium hydroxide solution.

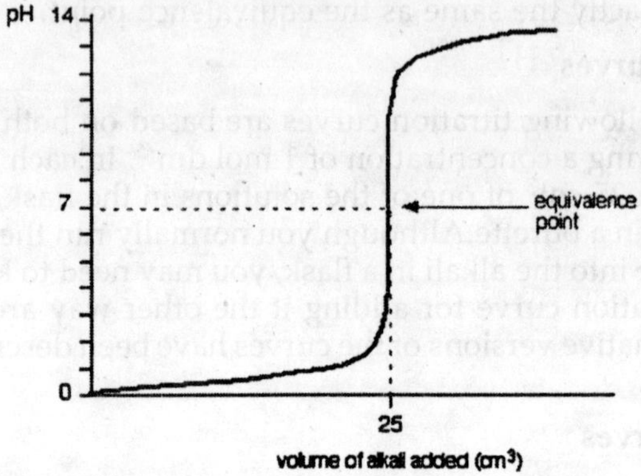

Again, the pH doesn't change very much until you get close to the equivalence point. Then it surges upwards very steeply. This time we are going to use hydrochloric acid as the strong acid and ammonia solution as the weak base.

$$NH_{3(aq)} + HCl_{(aq)} \longrightarrow NH_4Cl_{(aq)}$$

Running acid into the alkali

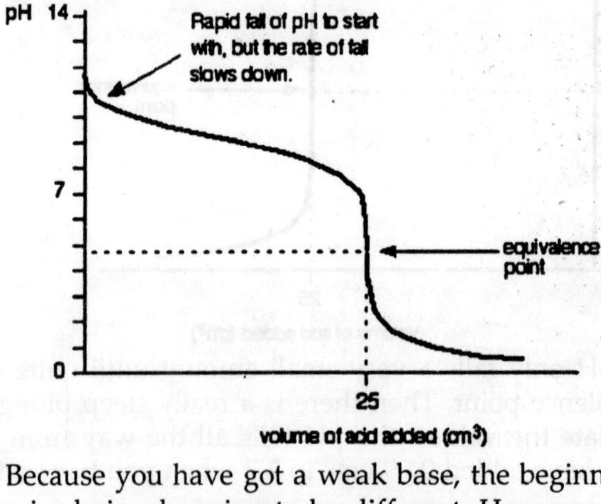

Because you have got a weak base, the beginning of the curve is obviously going to be different. However, once you have got an excess of acid, the curve is essentially the same as before.At the very beginning of the curve, the pH starts by falling quite quickly as the acid is added, but the curve very soon gets less steep. This is because a buffer solution is being set up - composed of the excess ammonia and the ammonium chloride being formed.The equivalence point is now somewhat acidic (a bit less than pH 5), because pure ammonium chloride isn't neutral. However, the equivalence point still falls on the steepest bit of the curve. That will turn out to be important in choosing a suitable indicator for the titration.

Running Alkali into the Acid

At the beginning of this titration, you have an excess of hydrochloric acid. The shape of the curve will be the same as when you had an excess of acid at the start of a titration running sodium hydroxide solution into the acid.It is only after the equivalence point that things become different.A buffer solution is formed containing excess ammonia and ammonium chloride. This resists any large increase in pH - not that you would expect a very large increase anyway, because ammonia is only a weak base.

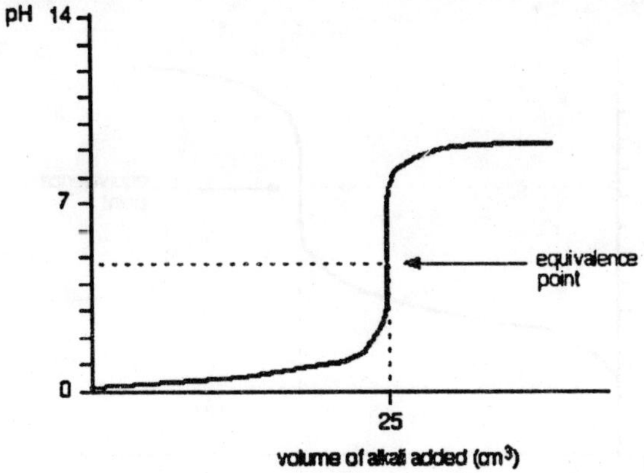

We'll take ethanoic acid and sodium hydroxide as typical of a weak acid and a strong base.

$$CH_3COOH_{(aq)} + NaOH_{(aq)} \longrightarrow CH_3COONa_{(aq)} + H_2O_{(l)}$$

Running acid into the alkali

For the first part of the graph, you have an excess of sodium hydroxide. The curve will be exactly the same as when you add hydrochloric acid to sodium hydroxide. Once the acid is in excess, there will be a difference.

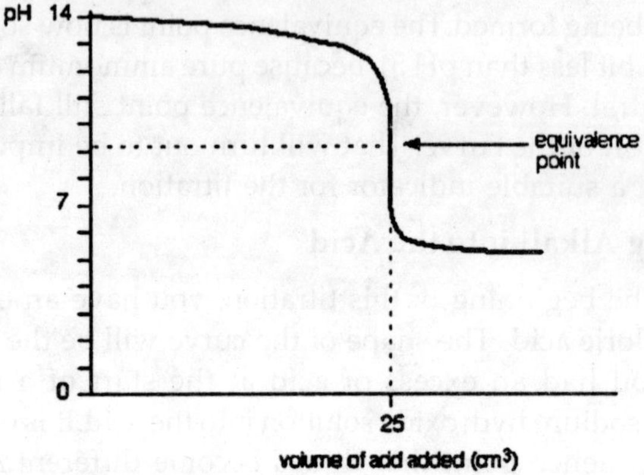

Past the equivalence point you have a buffer solution containing sodium ethanoate and ethanoic acid. This resists any large fall in pH.

Running alkali into the acid

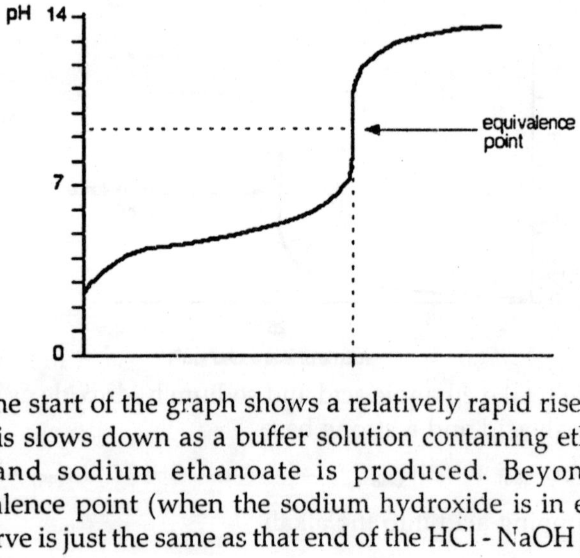

The start of the graph shows a relatively rapid rise in pH but this slows down as a buffer solution containing ethanoic acid and sodium ethanoate is produced. Beyond the equivalence point (when the sodium hydroxide is in excess) the curve is just the same as that end of the HCl - NaOH graph.

The common example of this would be ethanoic acid and ammonia.

$$CH_3COOH_{(aq)} + NH_{3(aq)} \longrightarrow CH_3COONH_{4(aq)}$$.

It so happens that these two are both about equally weak - in that case, the equivalence point is approximately pH 7.

Running acid into the alkali

This is really just a combination of graphs you have already seen. Up to the equivalence point it is similar to the ammonia - HCl case. After the equivalence point it is like the end of the ethanoic acid - NaOH curve.

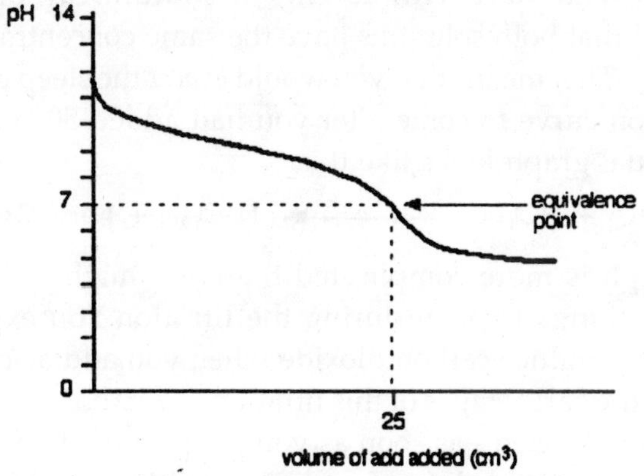

Notice that there isn't any steep bit on this graph. Instead, there is just what is known as a "point of inflexion". That lack of a steep bit means that it is difficult to do a titration of a weak acid against a weak base.

The way you normally carry out a titration involves adding the acid to the alkali.

The overall equation for the reaction between sodium carbonate solution and dilute hydrochloric acid is:

$$Na_2CO_{3(aq)} + 2HCl_{(aq)} \longrightarrow 2NaCl_{(aq)} + CO_{2(g)} + H_2O_{l}$$

If you had the two solutions of the same concentration, you would have to use twice the volume of hydrochloric acid to reach the equivalence point - because of the 1 : 2 ratio in the equation.

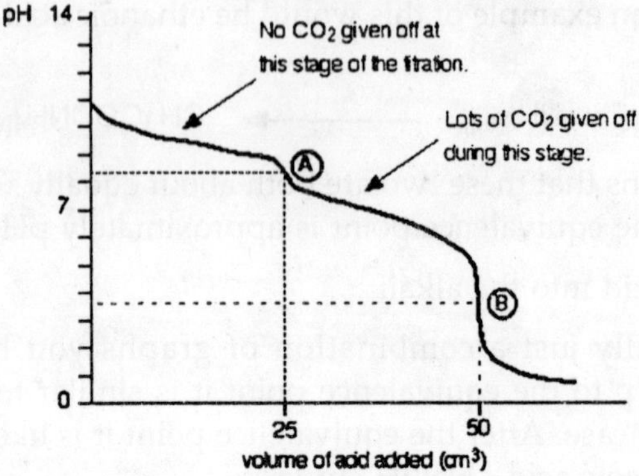

Suppose you start with 25 cm^3 of sodium carbonate solution, and that both solutions have the same concentration of 1 mol dm^{-3}. That means that you would *expect* the steep drop in the titration curve to come after you had added 50 cm^3 of acid.The actual graph looks like this:

$$Na_2CO_{3(aq)} + HCl_{(aq)} \longrightarrow NaCl_{(aq)} + NaHCO_{3(aq)}$$

The graph is more complicated than you might think - and curious things happen during the titration.You expect carbonates to produce carbon dioxide when you add acids to them, but in the early stages of this titration, no carbon dioxide is given off at all.Then - as soon as you get past the half-way point in the titration - lots of carbon dioxide is suddenly released.The graph is showing two end points - one at a pH of 8.3 (little more than a point of inflexion), and a second at about pH 3.7. The reaction is obviously happening in two distinct parts.In the first part, complete at A in the diagram, the sodium carbonate is reacting with the acid to produce sodium hydrogencarbonate:

$$NaHCO_{3(aq)} + HCl_{(aq)} \longrightarrow NaCl_{(aq)} + CO_{2(g)} + H_2O_{(l)}$$

The reaction doesn't produce any carbon dioxide.In the second part, the sodium hydrogencarbonate produced goes on to react with more acid - giving off lots of CO_2.

That reaction is finished at B on the graph.It is possible to pick up both of these end points by careful choice of indicator.

That is explained on the separate page on indicators.Adding sodium hydroxide solution to dilute ethanedioic acid

Ethanedioic acid was previously known as oxalic acid. It is a *diprotic acid,* which means that it can give away 2 protons (hydrogen ions) to a base. Something which can only give away one (like HCl) is known as a monoprotic acid.

The reaction with sodium hydroxide takes place in two stages because one of the hydrogens is easier to remove than the other. The two successive reactions are:

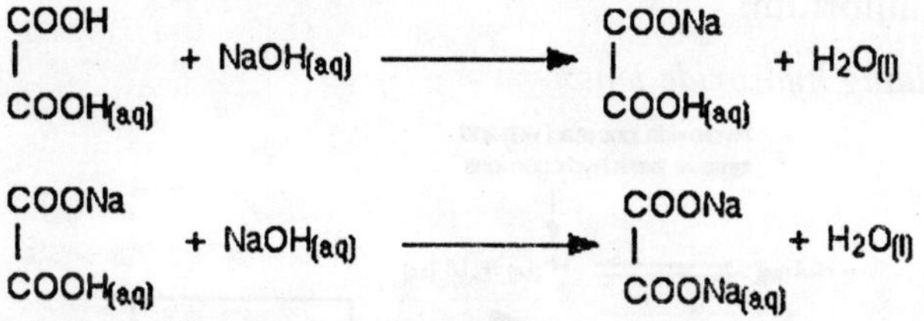

If you run sodium hydroxide solution into ethanedioic acid solution, the pH curve shows the end points for both of these reactions.

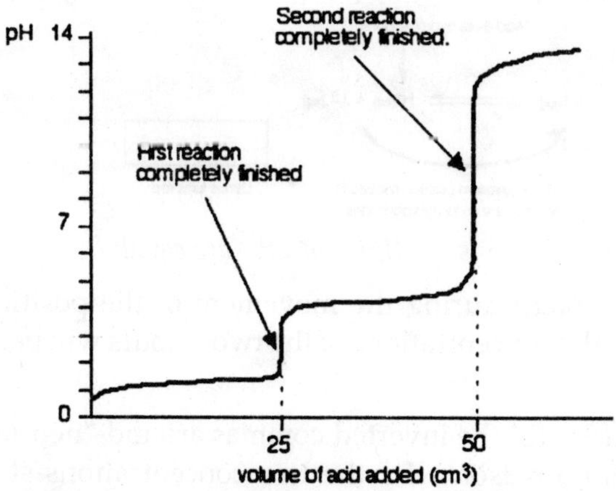

The curve is for the reaction between sodium hydroxide and ethanedioic acid solutions of equal concentrations.

ACID-BASE INDICATORS

Litmus is a weak acid. It has a seriously complicated molecule which we will simplify to HLit. The "H" is the proton which can be given away to something else. The "Lit" is the rest of the weak acid molecule.There will be an equilibrium established when this acid dissolves in water. Taking the simplified version of this equilibrium:

$$HLit_{(aq)} \rightleftharpoons H^+_{(aq)} + Lit^-_{(aq)}$$

The un-ionised litmus is red, whereas the ion is blue.Now use Le Chatelier's Principle to work out what would happen if you added hydroxide ions or some more hydrogen ions to this equilibrium.

Adding hydroxide ions:

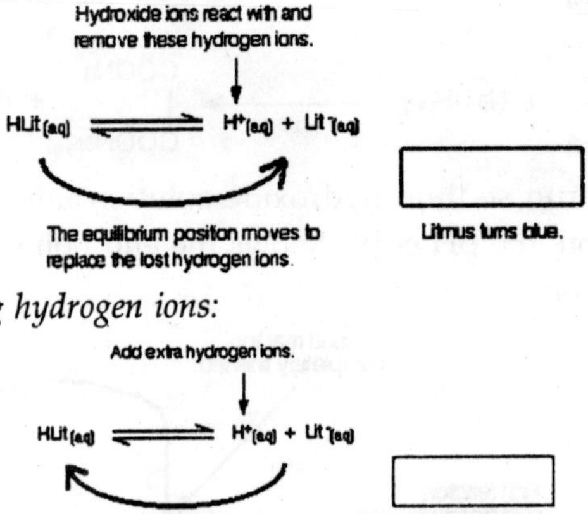

Adding hydrogen ions:

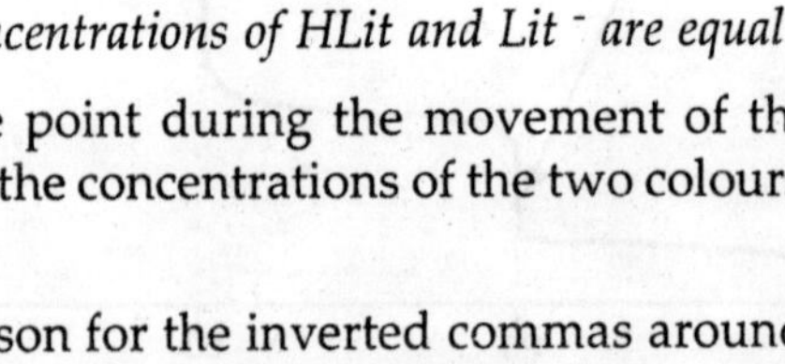

If the concentrations of HLit and Lit ⁻ are equal:

At some point during the movement of the position of equilibrium, the concentrations of the two colours will become equal.

The reason for the inverted commas around "neutral" is that there is no reason why the two concentrations should become equal at pH 7. For litmus, it so happens that the 50 / 50 colour does occur at close to pH 7 - that's why litmus is commonly used to test for acids and alkalis. As you will see below, that isn't true for other indicators.

Methyl Orange

Methyl orange is one of the indicators commonly used in titrations. In an alkaline solution, methyl orange is yellow and the structure is:

the yellow form of methyl orange

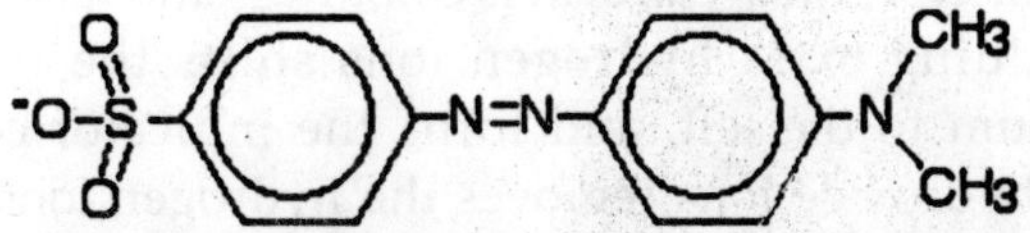

Now, you might think that when you add an acid, the hydrogen ion would be picked up by the negatively charged oxygen. That's the obvious place for it to go.In fact, the hydrogen ion attaches to one of the nitrogens in the nitrogen-nitrogen double bond to give a structure which might be drawn like this:

the red form of methyl orange

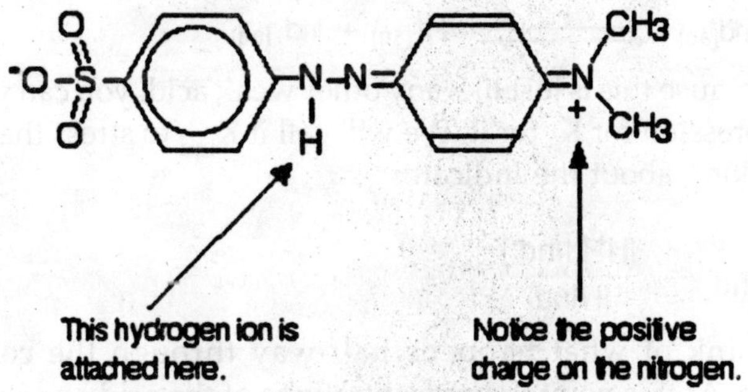

You have the same sort of equilibrium between the two forms of methyl orange as in the litmus case - but the colours are different.

$$H - Meor_{(aq)} \rightleftharpoons H^{+}_{(aq)} + Meor^{-}_{(aq)}$$

You should be able to work out for yourself why the colour changes when you add an acid or an alkali. The explanation is identical to the litmus case - all that differs are the colours.

In the methyl orange case, the half-way stage where the mixture of red and yellow produces an orange colour happens at pH 3.7 - nowhere near neutral. *Phenolphthalein*

Phenolphthalein is another commonly used indicator for titrations, and is another weak acid.

$$H - phph_{(aq)} \rightleftharpoons H^{+}_{(aq)} + phph^{-}_{(aq)}$$

In this case, the weak acid is colourless and its ion is bright pink. Adding extra hydrogen ions shifts the position of equilibrium to the left, and turns the indicator colourless. Adding hydroxide ions removes the hydrogen ions from the equilibrium which tips to the right to replace them - turning the indicator pink.The half-way stage happens at pH 9.3. Since a mixture of pink and colourless is simply a paler pink.

The pH range of indicators

Think about a general indicator, HInd - where "Ind" is all the rest of the indicator apart from the hydrogen ion which is given away:

$$HInd_{(aq)} \rightleftharpoons H^{+}_{(aq)} + Ind^{-}_{(aq)}$$

Because this is just like any other weak acid, you can write an expression for K_a for it. We will call it K_{ind} to stress that we are talking about the indicator.

$$K_{ind} = \frac{[H^{+}][Ind^{-}]}{[HInd]}$$

Think of what happens half-way through the colour change. At this point the concentrations of the acid and its ion are equal. In that case, they will cancel out of the K_{ind} expression.

$$K_{ind} = [H^{+}]$$

You can use this to work out what the pH is at this half-way point. If you re-arrange the last equation so that the hydrogen ion concentration is on the left-hand side, and then convert to pH and pK_{ind}, you get:

$$[H^+] = k_{ind}$$

$$pH = pk_{ind}$$

That means that the end point for the indicator depends entirely on what its pK_{ind} value is. For the indicators we've looked at above, these are:

indicator	pK_{ind}
litmus	6.5
methyl orange	3.7
phenolphthalein	9.3

The pH range of Indicators

Indicators don't change colour sharply at one particular pH (given by their pK_{ind}). Instead, they change over a narrow range of pH.Assume the equilibrium is firmly to one side, but now you add something to start to shift it. As the equilibrium shifts, you will start to get more and more of the second colour formed, and at some point the eye will start to detect it.

For example, suppose you had methyl orange in an alkaline solution so that the dominant colour was yellow. Now start to add acid so that the equilibrium begins to shift.At some point there will be enough of the red form of the methyl orange present that the solution will begin to take on an orange tint. As you go on adding more acid, the red will eventually become so dominant that you can no longe see any yellow.There is a gradual smooth change from one colour to the other, taking place over a range of pH. As a rough "rule of thumb", the visible change takes place about 1 pH unit either side of the pK_{ind} value.The exact values for the three indicators we've looked at are:

Indicator	pK_{ind}	pH range
litmus	6.5	5 - 8
methyl orange	3.7	3.1 - 4.4
phenolphthalein	9.3	8.3 - 10.0

The litmus colour change happens over an unusually wide range, but it is useful for detecting acids and alkalis in the lab because it changes colour around pH 7. Methyl orange or phenolphthalein would be less useful.

For example, methyl orange would be yellow in any solution with a pH greater than 4.4. It couldn't distinguish between a weak acid with a pH of 5 or a strong alkali with a pH of 14.

Choosing Indicators for Titrations

Remember that the equivalence point of a titration is where you have mixed the two substances in exactly equation proportions. You obviously need to choose an indicator which changes colour as close as possible to that equivalence point. That varies from titration to titration.

Strong Acid v Strong Base

The next diagram shows the pH curve for adding a strong acid to a strong base. Superimposed on it are the pH ranges for methyl orange and phenolphthalein.

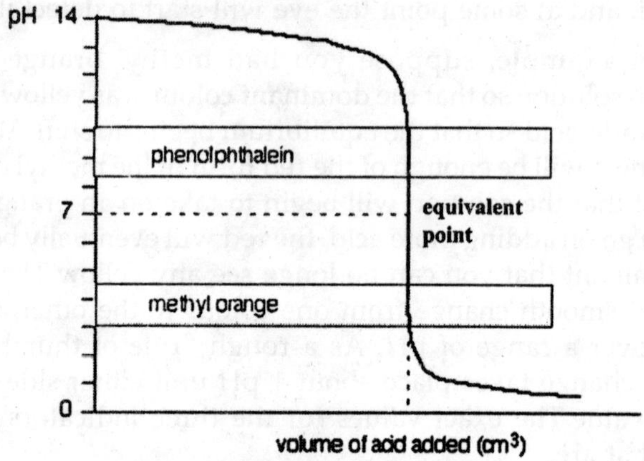

Neither indicator changes colour at the equivalence point.However, the graph is so steep at that point that there will be virtually no difference in the volume of acid added whichever indicator you choose. However, it would make sense to titrate to the best possible colour with each indicator.If you use phenolphthalein, you would titrate until it just becomes colourless (at pH 8.3) because that is as close as you can get to the equivalence point.On the other hand, using methyl orange, you would titrate until there is the very first trace of orange in the solution. If the solution becomes red,

Strong acid v Weak Base

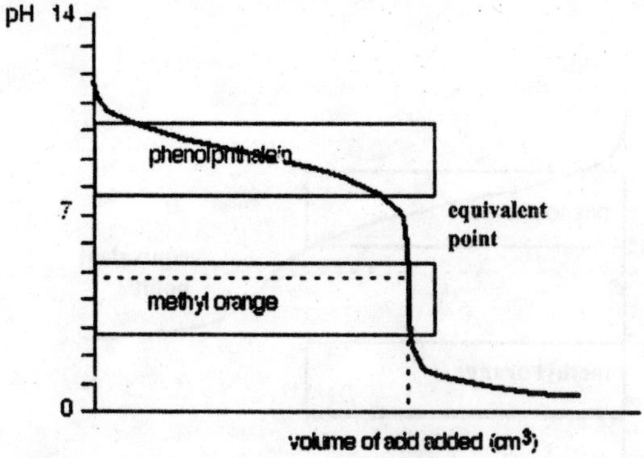

This time it is obvious that phenolphthalein would be completely useless. However, methyl orange starts to change from yellow towards orange very close to the equivalence point.You have to choose an indicator which changes colour on the steep bit of the curve.

Weak Acid v Strong Base

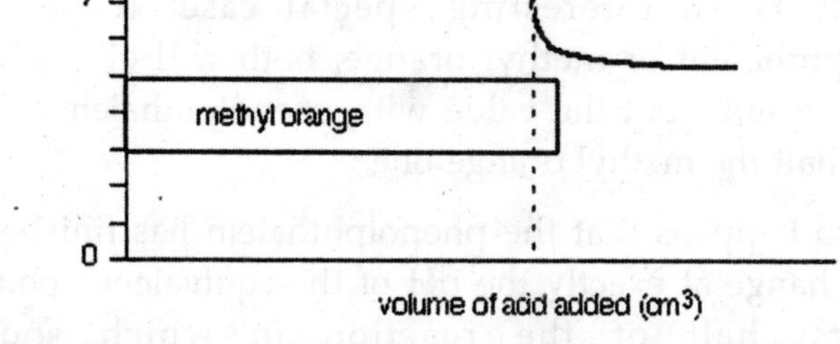

This time, the methyl orange is hopeless! However, the phenolphthalein changes colour exactly where you want it to.

Weak Acid v Weak Base

The curve is for a case where the acid and base are both equally weak - for example, ethanoic acid and ammonia solution. In other cases, the equivalence point will be at some other pH.

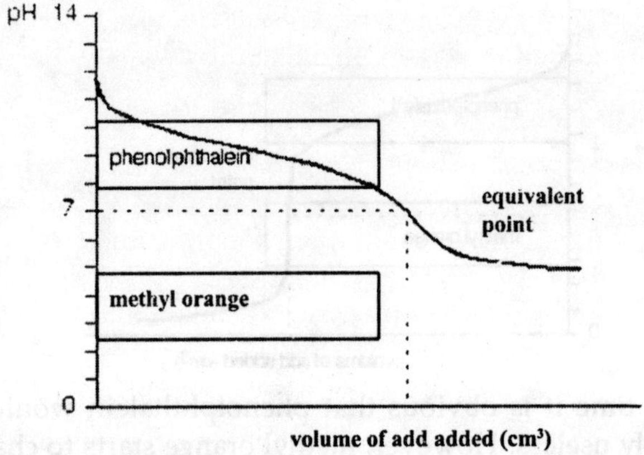

Neither indicator is any use. Phenolphthalein will have finished changing well before the equivalence point, and methyl orange falls off the graph altogether.It may be possible to find an indicator which starts to change or finishes changing at the equivalence point, but because the pH of the equivalence point will be different from case to case,On the whole, you would never titrate a weak acid and a weak base in the presence of an indicator.

Sodium carbonate solution and dilute hydrochloric acid

This is an interesting special case. If you use phenolphthalein or methyl orange, both will give a valid titration result - but the value with phenolphthalein will be exactly half the methyl orange one.

It so happens that the phenolphthalein has finished its colour change at exactly the pH of the equivalence point of the first half of the reaction in which sodium hydrogencarbonate is produced.

$$Na_2CO_{3(aq)} + HCl_{(aq)} \longrightarrow NaCl_{(aq)} + NaHCO_{3(aq)}$$

The methyl orange changes colour at exactly the pH of the equivalence point of the second stage of the reaction.

$$NaHCO_{3(aq)} + HCl_{(aq)} \longrightarrow NaCl_{(aq)} + CO_{2(g)} + H_2O_{(l)}$$

BUFFER SOLUTIONS

A buffer solution is one which resists changes in pH when small quantities of an acid or an alkali are added to it.

Acidic Buffer Solutions

An acidic buffer solution is simply one which has a pH less than 7. Acidic buffer solutions are commonly made from a weak acid and one of its salts - often a sodium salt.A common example would be a mixture of ethanoic acid and sodium ethanoate in solution. In this case, if the solution contained equal molar concentrations of both the acid and the salt, it would have a pH of 4.76. It wouldn't matter what the concentrations were, as long as they were the same.You can change the pH of the buffer solution by changing the ratio of acid to salt, or by choosing a different acid and one of its salts.

Alkaline Buffer Solutions

An alkaline buffer solution has a pH greater than 7. Alkaline buffer solutions are commonly made from a weak base and one of its salts.A frequently used example is a mixture of ammonia solution and ammonium chloride solution. If these were mixed in equal molar proportions, the solution would have a pH of 9.25. Again, it doesn't matter what concentrations you choose as long as they are the same.A buffer solution has to contain things which will remove any hydrogen ions or hydroxide ions that you might add to it - otherwise the pH will change. Acidic and alkaline buffer solutions achieve this in different ways.

Acidic Buffer Solutions

We'll take a mixture of ethanoic acid and sodium ethanoate as typical.Ethanoic acid is a weak acid, and the position of this equilibrium will be well to the left:

$$CH_3COOH_{(aq)} \rightleftharpoons CH_3COO^-_{(aq)} + H^+_{(aq)}$$

Adding sodium ethanoate to this adds lots of extra ethanoate ions. According to Le Chatelier's Principle, that will tip the position of the equilibrium even further to the left.

The solution will therefore contain these important things:

- lots of un-ionised ethanoic acid;
- lots of ethanoate ions from the sodium ethanoate;
- enough hydrogen ions to make the solution acidic.

Other things (like water and sodium ions) which are present aren't important to the argument.

Adding an acid to this buffer solution

The buffer solution must remove most of the new hydrogen ions otherwise the pH would drop markedly.Hydrogen ions combine with the ethanoate ions to make ethanoic acid. Although the reaction is reversible, since the ethanoic acid is a weak acid, most of the new hydrogen ions are removed in this way.

$$CH_3COO^-_{(aq)} + H^+_{(aq)} \rightleftharpoons CH_3COOH_{(aq)}$$

Since most of the new hydrogen ions are removed, the pH won't change very much - but because of the equilibria involved, it *will* fall a little bit.

Adding an alkali to this buffer solution

Alkaline solutions contain hydroxide ions and the buffer solution removes most of these.This time the situation is a bit more complicated because there are *two* processes which can remove hydroxide ions.

Removal by reacting with ethanoic acid

The most likely acidic substance which a hydroxide ion is going to collide with is an ethanoic acid molecule. They will react to form ethanoate ions and water.

$$CH_3COOH_{(aq)} + OH^-_{(aq)} \rightleftharpoons CH_3COO^-_{(aq)} + H_2O_{(l)}$$

Because most of the new hydroxide ions are removed, the pH doesn't increase very much.

Removal of the hydroxide ions by reacting with hydrogen ions

Remember that there are some hydrogen ions present from the ionisation of the ethanoic acid.

$$CH_3COOH_{(aq)} \rightleftharpoons CH_3COO^-_{(aq)} + H^+_{(aq)}$$

Hydroxide ions can combine with these to make water. As soon as this happens, the equilibrium tips to replace them. This keeps on happening until most of the hydroxide ions are removed.

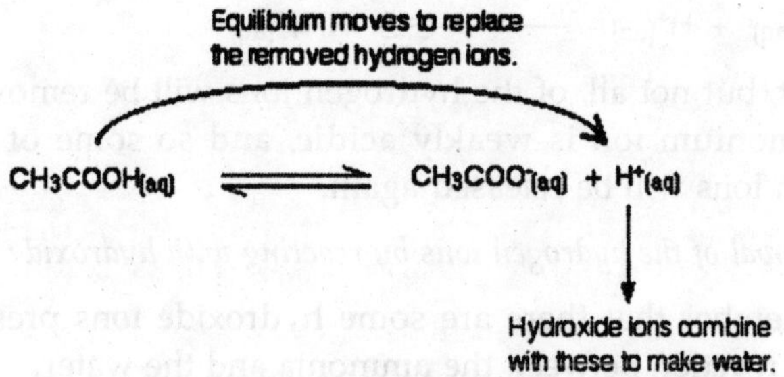

Again, because you have equilibria involved, not *all* of the hydroxide ions are removed - just most of them. The water formed re-ionises to a very small extent to give a few hydrogen ions and hydroxide ions.

Alkaline Buffer Solutions

We'll take a mixture of ammonia and ammonium chloride solutions as typical.Ammonia is a weak base, and the position of this equilibrium will be well to the left:

$$NH_{3(aq)} + H_2O_{(l)} \rightleftharpoons NH_4^+{}_{(aq)} + OH^-_{(aq)}$$

Adding ammonium chloride to this adds lots of extra ammonium ions. According to Le Chatelier's Principle, that will tip the position of the equilibrium even further to the left.The solution will therefore contain these important things:

- lots of unreacted ammonia;
- lots of ammonium ions from the ammonium chloride;
- enough hydroxide ions to make the solution alkaline.

Other things (like water and chloride ions) which are present aren't important to the argument.

Adding an acid to this buffer solution

There are *two* processes which can remove the hydrogen ions that you are adding.

Removal by reacting with ammonia

The most likely basic substance which a hydrogen ion is going to collide with is an ammonia molecule. They will react to form ammonium ions.

$$NH_{3(aq)} + H^+_{(aq)} \rightleftharpoons NH_4^+{}_{(aq)}$$

Most, but not all, of the hydrogen ions will be removed. The ammonium ion is weakly acidic, and so some of the hydrogen ions will be released again.

Removal of the hydrogen ions by reacting with hydroxide ions

Remember that there are some hydroxide ions present from the reaction between the ammonia and the water.

$$NH_{3(aq)} + H_2O_{(l)} \rightleftharpoons NH_4^+{}_{(aq)} + OH^-_{(aq)}$$

Hydrogen ions can combine with these hydroxide ions to make water. As soon as this happens, the equilibrium tips to replace the hydroxide ions. This keeps on happening until most of the hydrogen ions are removed.

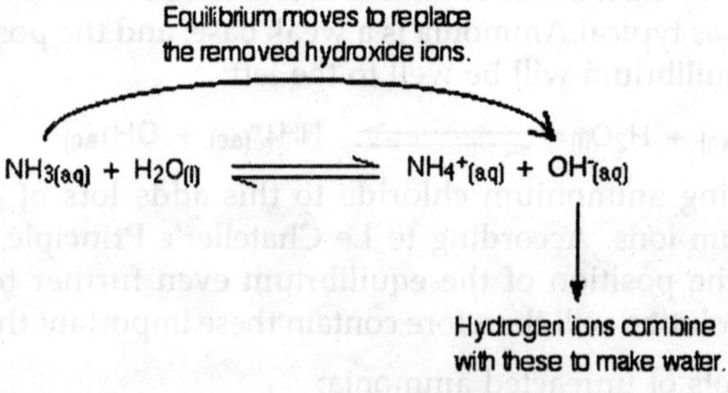

Again, because you have equilibria involved, not *all* of the hydrogen ions are removed - just most of them.

Adding an alkali to this buffer solution

The hydroxide ions from the alkali are removed by a simple reaction with ammonium ions.

$$NH_4^+{}_{(aq)} + OH^-{}_{(aq)} \rightleftharpoons NH_{3(aq)} + H_2O_{(l)}$$

Because the ammonia formed is a weak base, it can react with the water - and so the reaction is slightly reversible. That means that, again, most (but not all) of the the hydroxide ions are removed from the solution.

Chapter 8

Periodic Table

ALKALI METALS

The alkali metals are the first group of the periodic table of the elements. These metals are lithium (Li), sodium (Na), potassium (K), rubidium (Rb), caesium (Cs) and francium (Fr). They have got one outer electron and the connection to the atomic nucleus is very weak. They are the least noble substances. In the air they react very fast with oxygen and so they are covered in an oxide layer. Alkali metals are kept in liquid hydrocarbons like petroleum. The reactivity grows from lithium to caesium. They react suddenly with water, then they burn or can even explode. Because of the big radius of caesium his outer electron is very free, so caesium reacts so strong with oxygen that it begins to burn in the air. When these metals react with water the product is a caustic solution, like caustic soda solution (NaOH). They are very soft, too.

ALKALINE EARTH METALS

The alkaline earth metals are the second group of the periodic table of the elements. These metals are beryllium (Be), magnesium (Mg), calcium (Ca), strontium (Sr), barium (Ba) and radium (Ra). These elements have got two outer electrons and so they did not react as strong as the alkali metals, because also can make caustic solutions. These elements are much harder than the alkali metals. Magnesium is an element which do not only burns in oxygen, but in all substances which contain oxygen. So magnesium burns in water, too. Divers use magnesium flares to see something under water.

Boron Group

The boron group is the third group of the periodic table of the elements. These elements are boron (B), aluminium (Al), gallium (Ga), Indium (In) and thallium (Tl). Boron is a metalloid and there other four elements are metals. The most used element of this group is aluminium. They elements are reactively. When aluminium is used it is covered in an oxide layer.

CARBON SILICON GROUP

The carbon silicon group is the fourth group of the periodic table of the elements. These elements are carbon (C), silicon (Si), germanium (Ge), tin (Sn) and lead (Pb). Carbon is so noble that it is a nonmetal. Silicon and germanium are metalloids and tin and lead are metals. Lead is a very soft metal. Silicon is very often found in the compound silicon dioxide (quartz). Silicon and germanium are very important semiconductor and so they are very important for elektronics.

PNICTOGENS

The pnictogens are the fifth group of the periodic table of the elements. These elements are nitrogen (N), phosphorus (P), arsenic (As), antimony (Sb) and bismuth (Bi). Nitrogen and phosphorus are nonmetals. Arsenic and antimony are metalloids and bismuth is a metal. 80% of the earth atmosphere are nitrogen, but in molecules (N_2).

CHALCOGENS

The chalcogens are the sixth group of the periodic table of the elements. These elements are oxygen (O), sulfur (S), selenium (Se), tellurium (Te) and polonium (Po). Oxygen and sulfur and nonmetals. Selenium and tellurium are metalloids and polonium is a metal.

Halogens

The halogens are the seventh group of the periodic table of the elements. These elements are fluorine (F), chlorine (Cl), bromine (Br), iodine (I) and astatine (At). Fluorine, chlorine, bromine and iodine are nonmetals. Astatine is a metalloid. The halogens only need one electron to reach the noble gas configuration and so they are very reactively. Fluor is the most reactively element of all elements. Many substances (for example wood) begin to burn, because of the contact with fluorine. Fluorine is a colorless gas. The other halogens are very dangerous, because they are very reactively and very corrosive, too. Clorine was used as a strong corrosive poison in the First World War. It is yellow-green. Bromine is a dark liquid at room temperature and like all other halogens a very corrosive substance.

NOBLE GASES

The noble gases are the eigthth group of the periodic table of the elements. These elements are helium (He), neon (Ne), argon (Ar), krypton (Kr), xenon (Xe) and radon (Rn). They are all nonmetals and like the name it says they are gases. They have got 8 outer electron (helium: 2 outer electrons) and they nearly do not react at all. They are the most noble elements.

TRANSITION METALS

We have not got mentioned the most elements of the periodic table of the elements. These are the transition metals. They metals where the other atomic shells are filled. Famous metals are for example gold (Au), silver (Ag), iron (Fe), copper (Cu), mercury (Hg) and many others. The lanthanoids (rare-earths) and the actionoids belong also to the transition metals. Famous actinoids are the both radioactive elements uranium (U) and plutonium (Pu).

Hydrogen

Hydrogen is a chemical element represented by the symbol H and an atomic number of 1. At standard temperature and pressure it is a colorless, odorless, nonmetallic, tasteless, highly flammable diatomic gas (H_2). With an atomic mass of 1.00794 g/mol, hydrogen is the lightest element.

Hydrogen is the most abundant of the chemical elements, constituting roughly 75% of the universe's elemental mass.

Stars in the main sequence are mainly composed of hydrogen in its plasma state. Elemental hydrogen is relatively rare on Earth, and is industrially produced from hydrocarbons such as methane, after which most elemental hydrogen is used "captively" (meaning locally at the production site), with the largest markets about equally divided between fossil fuel upgrading (e.g., hydrocracking) and in ammonia production (mostly for the fertilizer market). Hydrogen may be produced from water using the process of electrolysis, but this process is presently significantly more expensive commercially than hydrogen production from natural gas.

The most common naturally occurring isotope of hydrogen, known as protium, has a single proton and no neutrons. In ionic compounds it can take on either a positive charge (becoming a cation composed of a bare proton) or a negative charge (becoming an anion known as a hydride). Hydrogen can form compounds with most elements and is present in water and most organic compounds. It plays a particularly important role in acid-base chemistry, in which many reactions involve the exchange of protons between soluble molecules. As the only neutral atom for which the Schrödinger equation can be solved analytically, study of the energetics and bonding of the hydrogen atom has played a key role in the development of quantum mechanics.

NOMENCLATURE

Hydrogen, Latin: 'hydrogenium', is from Ancient Greek (*hydor*): "water" and (*genes*): "forming". Ancient Greek (*geinomai*): "to beget or sire")

The word "hydrogen" has several different meanings;

1. The *name of an element*.
2. An *atom*, sometimes called "H dot", that is abundant in space but essentially absent on Earth, because it dimerizes.
3. A *diatomic molecule* that occurs naturally in trace amounts in the Earth's atmosphere; chemists increasingly refer to H_2 as *dihydrogen*,or *hydrogen*

molecule, to distinguish this molecule from atomic hydrogen and hydrogen found in other compounds.

4. The atomic *constituent* within all organic compounds, water, and many other chemical compounds.

The *elemental* forms of hydrogen should not be confused with hydrogen as it appears in chemical compounds.

HISTORY

Discovery of H_2

Hydrogen gas, H_2, was first artificially produced and formally described by T. Von Hohenheim via the mixing of metals with strong acids. He was unaware that the flammable gas produced by this chemical reaction was a new chemical element. In 1671, Robert Boyle rediscovered and described the reaction between iron filings and dilute acids, which results in the production of hydrogen gas.In 1766, Henry Cavendish was the first to recognize hydrogen gas as a discrete substance, by identifying the gas from a metal-acid reaction as "inflammable air" and further finding that the gas produces water when burned. Cavendish had stumbled on hydrogen when experimenting with acids and mercury. Although he wrongly assumed that hydrogen was a liberated component of the mercury rather than the acid, he was still able to accurately describe several key properties of hydrogen. He is usually given credit for its discovery as an element. In 1783, Antoine Lavoisier gave the element the name of hydrogen when he (with Laplace) reproduced Cavendish's finding that water is produced when hydrogen is burned. Lavoisier's name for the gas won out.

One of the first uses of H_2 was for balloons, and later airships. The H_2 was obtained by reacting sulphuric acid and metallic iron. Infamously, H_2 was used in the Hindenburg airship that was destroyed in a midair fire. The highly flammable hydrogen (H_2) was later replaced for airships and most balloons by the unreactive helium (He).

Role in History of Quantum Theory

Because of its relatively simple atomic structure, consisting only of a proton and an electron, the hydrogen atom, together with the spectrum of light produced from it or absorbed by it, has been central to the development of the theory of atomic structure. Furthermore, the corresponding simplicity of the hydrogen molecule and the corresponding cation H_2^+ allowed fuller understanding of the nature of the chemical bond, which followed shortly after the quantum mechanical treatment of the hydrogen atom had been developed in the mid-1920s.

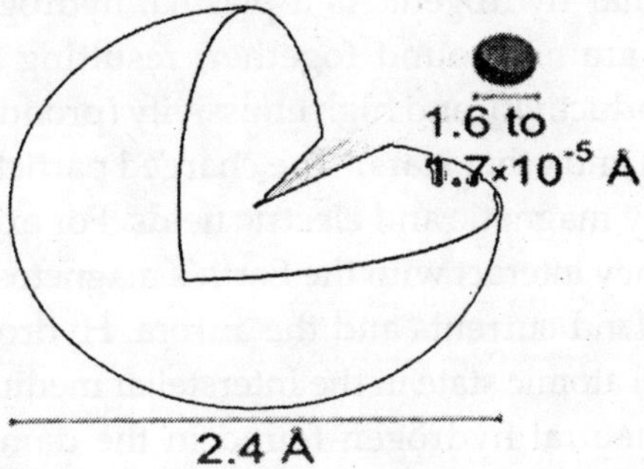

One of the first quantum effects to be explicitly noticed (but not understood at the time) was a Maxwell observation involving hydrogen, half a century before full quantum mechanical theory arrived. Maxwell observed that the specific heat capacity of H_2 unaccountably departs from that of a diatomic gas below room temperature and begins to increasingly resemble that of a monatomic gas at cryogenic temperatures. According to quantum theory, this behavior arises from the spacing of the (quantized) rotational energy levels, which are particularly wide-spaced in H_2 because of its low mass. These widely spaced levels inhibit equal partition of heat energy into rotational motion in hydrogen at low temperatures. Diatomic gases composed of heavier atoms do not have such widely spaced levels and do not exhibit the same effect.

Natural Occurrence

A giant region of ionized hydrogen in the Triangulum Galaxy.Hydrogen is the most abundant element in the universe, making up 75% of normal matter by mass and over 90% by number of atoms. This element is found in great abundance in stars and gas giant planets. Molecular clouds of H_2 are associated with star formation. Hydrogen plays a vital role in powering stars through proton-proton reaction nuclear fusion.

Throughout the universe, hydrogen is mostly found in the atomic and plasma states whose properties are quite different from molecular hydrogen. As a plasma, hydrogen's electron and proton are not bound together, resulting in very high electrical conductivity and high emissivity (producing the light from the sun and other stars). The charged particles are highly influenced by magnetic and electric fields. For example, in the solar wind they interact with the Earth's magnetosphere giving rise to Birkeland currents and the aurora. Hydrogen is found in the neutral atomic state in the Interstellar medium. The large amount of neutral hydrogen found in the damped Lyman-alpha systems is thought to dominate the cosmological baryonic density of the Universe up to redshift $z=4$.

Under ordinary conditions on Earth, elemental hydrogen exists as the diatomic gas, H_2. However, hydrogen gas is very rare in the Earth's atmosphere (1 ppm by volume) because of its light weight, which enables it to escape from Earth's gravity more easily than heavier gases. Although H atoms and H_2 molecules are abundant in interstellar space, they are difficult to generate, concentrate, and purify on Earth. Still, hydrogen is the third most abundant element on the Earth's surface.Most of the Earth's hydrogen is in the form of chemical compounds such as hydrocarbons and water. Hydrogen gas is produced by some bacteria and algae and is a natural component of flatus. Methane is a hydrogen source of increasing importance.

The Hydrogen Atom

Electron Energy Levels

Depiction of a hydrogen atom showing the diameter as about twice the Bohr model radius.(Image not to scale)

The ground state energy level of the electron in a hydrogen atom is 13.6 eV, which is equivalent to an ultraviolet photon of roughly 92 nm.

The energy levels of hydrogen can be calculated fairly accurately using the Bohr model of the atom, which conceptualizes the electron as "orbiting" the proton in analogy to the Earth's orbit of the sun. However, the electromagnetic force attracts electrons and protons to one another, while planets and celestial objects are attracted to each other by gravity. Because of the discretization of angular momentum postulated in early quantum mechanics by Bohr, the electron in the Bohr model can only occupy certain allowed distances from the proton, and therefore only certain allowed energies. A more accurate description of the hydrogen atom comes from a purely quantum mechanical treatment that uses the Schrödinger equation or the equivalent Feynman path integral formulation to calculate the probability density of the electron around the proton. Treating the electron as a matter wave reproduces chemical results such as shape of the hydrogen atom more naturally than the particle-based Bohr model, although the energy and spectral results are the same. Modeling the system fully using the reduced mass of nucleus and electron (as one would do in the two-body problem in celestial mechanics) yields an even better formula for the hydrogen spectra, and also the correct spectral shifts for the isotopes deuterium and tritium. Very small adjustments in energy levels in the hydrogen atom, which correspond to actual spectral effects, may be determined by using a full quantum mechanical theory which corrects for the effects of special relativity, and by accounting for quantum effects arising from production of virtual particles in the vacuum and as a result of electric fields.

In hydrogen gas, the electronic ground state energy level is split into hyperfine structure levels because of magnetic effects of the quantum mechanical spin of the electron and proton. The energy of the atom when the proton and electron spins are aligned is higher than when they are not aligned. The transition between these two states can occur through emission of a photon through a magnetic dipole transition. Radio telescopes can detect the radiation produced in this process, which is used to map the distribution of hydrogen in the galaxy.

Isotopes

Protium, the most common isotope of hydrogen, has one proton and one electron. Unique among all stable isotopes, it has no neutrons.

Hydrogen has three naturally occurring isotopes, denoted ^{1}H, ^{2}H, and ^{3}H. Other, highly unstable nuclei (^{4}H to ^{7}H) have been synthesized in the laboratory but not observed in nature.

- ^{1}H is the most common hydrogen isotope with an abundance of more than 99.98%. Because the nucleus of this isotope consists of only a single proton, it is given the descriptive but rarely used formal name *protium*.
- ^{2}H, the other stable hydrogen isotope, is known as *deuterium* and contains one proton and one neutron in its nucleus. Deuterium comprises 0.0026 – 0.0184% (by mole-fraction or atom-fraction) of hydrogen samples on Earth, with the lower number tending to be found in samples of hydrogen gas and the higher enrichments (0.015% or 150 ppm) typical of ocean water. Deuterium is not radioactive, and does not represent a significant toxicity hazard. Water enriched in molecules that include deuterium instead of normal hydrogen is called heavy water. Deuterium and its compounds are used as a non-radioactive label in chemical experiments and in solvents for ^{1}H-NMR spectroscopy. Heavy water is used as a neutron

moderator and coolant for nuclear reactors. Deuterium is also a potential fuel for commercial nuclear fusion.

- ^3H is known as *tritium* and contains one proton and two neutrons in its nucleus. It is radioactive, decaying into Helium-3 through beta decay with a half-life of 12.32 years. Small amounts of tritium occur naturally because of the interaction of cosmic rays with atmospheric gases; tritium has also been released during nuclear weapons tests. It is used in nuclear fusion reactions, as a tracer in isotope geochemistry, and specialized in self-powered lighting devices. Tritium was once routinely used in chemical and biological labeling experiments as a radiolabel.

Hydrogen is the only element that has different names for its isotopes in common use today. (During the early study of radioactivity, various heavy radioactive isotopes were given names, but such names are no longer used). The symbols D and T (instead of ^2H and ^3H) are sometimes used for deuterium and tritium, but the corresponding symbol P is already in use for phosphorus and thus is not available for protium. IUPAC states that while this use is common it is not preferred.

Elemental Molecular Forms

First tracks observed in liquid hydrogen bubble chamber at the Bevatron.

There are two different types of diatomic hydrogen molecules that differ by the relative spin of their nuclei.In the orthohydrogen form, the spins of the two protons are parallel and form a triplet state; in the parahydrogen form the spins are antiparallel and form a singlet. At standard temperature and pressure, hydrogen gas contains about 25% of the para form and 75% of the ortho form, also known as the "normal form". The equilibrium ratio of orthohydrogen to parahydrogen depends on temperature, but since the ortho form is an excited state and has a higher energy than the para form, it is unstable and cannot be purified. At very low temperatures, the equilibrium state is composed almost exclusively of the para form. The physical properties of pure parahydrogen differ slightly from those of the normal form.The ortho/para distinction also occurs in other hydrogen-containing molecules or functional groups, such as water and methylene.

The uncatalyzed interconversion between para and ortho H_2 increases with increasing temperature; thus rapidly condensed H_2 contains large quantities of the high-energy ortho form that convert to the para form very slowly.The ortho/para ratio in condensed H_2 is an important consideration in the preparation and storage of liquid hydrogen: the conversion from ortho to para is exothermic and produces enough heat to evaporate the hydrogen liquid, leading to loss of the liquefied material. Catalysts for the ortho-para interconversion, such as iron compounds, are used during hydrogen cooling.

A molecular form called protonated molecular hydrogen, or H_3^+, is found in the interstellar medium (ISM), where it is generated by ionization of molecular Hydrogen from cosmic rays. It has also been observed in the upper atmosphere of the planet Jupiter. This molecule is relatively stable in the environment of outer space due to the low temperature and density. H_3^+ is one of the most abundant ions in the Universe, and it plays a notable role in the chemistry of the interstellar medium.

Chemical and Physical Properties

The solubility and adsorption characteristics of hydrogen with various metals are very important in metallurgy and in developing safe ways to store it for use as a fuel. Hydrogen is highly soluble in many compounds composed of rare earth metals and transition metals and can be dissolved in both crystalline and amorphous metals. Hydrogen solubility in metals is influenced by local distortions or impurities in the metal crystal lattice.

Combustion

Hydrogen can combust rapidly in air. It burned rapidly in the *Hindenburg* disaster on May 6, 1937.Hydrogen gas is highly flammable and will burn at concentrations as low as 4% H_2 in air. The enthalpy of combustion for hydrogen is – 286 kJ/mol; it combusts according to the following balanced equation.

$$2\ H_2(g) + O_2(g)\ '!\ 2\ H_2O(l) + 572\ kJ/mol$$

When mixed with oxygen across a wide range of proportions, hydrogen explodes upon ignition. Hydrogen burns violently in air. Pure hydrogen-oxygen flames are nearly invisible to the naked eye, as illustrated by the faintness of flame from the main Space Shuttle engines. Thus it is difficult to visually detect if a hydrogen leak is burning. The Hindenburg zeppelin flames seen in the adjacent picture are hydrogen flames colored with material from the covering skin of the zeppelin which contained carbon and pyrophoric aluminium powder, as well as other combustible materials. (Regardless of the cause of this fire, this was clearly primarily a hydrogen fire since skin of the Zeppelin alone would have taken many hours to burn). Another characteristic of hydrogen fires is that the flames tend to ascend rapidly with the gas in air, as illustrated by the Hindenberg flames, causing less damage than hydrocarbon fires. For example, two-thirds of the Hindenburg passengers survived the hydrogen fire, and many of the deaths which occurred were from falling or from gasoline burns.

H_2 reacts directly with other oxidizing elements. A violent and spontaneous reaction can occur at room temperature with chlorine and fluorine, forming the corresponding hydrogen halides: hydrogen chloride and hydrogen fluoride.

Compounds

Covalent and Organic Compounds

While H_2 is not very reactive under standard conditions, it does form compounds with most elements. Millions of hydrocarbons are known, but they are not formed by the direct reaction of elementary hydrogen and carbon (although synthesis gas production followed by the Fischer-Tropsch process to make hydrocarbons comes close to being an exception, as this begins with coal and the elemental hydrogen is generated in situ). Hydrogen can form compounds with elements that are more electronegative, such as halogens (e.g., F, Cl, Br, I) and chalcogens (O, S, Se); in these compounds hydrogen takes on a partial positive charge.

When bonded to fluorine, oxygen, or nitrogen, hydrogen can participate in a form of strong noncovalent bonding called hydrogen bonding, which is critical to the stability of many biological molecules. Hydrogen also forms compounds with less electronegative elements, such as the metals and metalloids, in which it takes on a partial negative charge. These compounds are often known as hydrides.

Hydrogen forms a vast array of compounds with carbon. Because of their general association with living things, these compounds came to be called organic compounds; the study of their properties is known as organic chemistry and their study in the context of living organisms is known as biochemistry. By some definitions, "organic" compounds are only required to contain carbon (as a classic historical example, urea). However, most of them also contain hydrogen, and since it is the carbon-hydrogen bond which gives this class of compounds most of its particular chemical characteristics, carbon-hydrogen bonds are required in some definitions of the word "organic" in chemistry. (This latter definition is not perfect, however, as in this definition urea would *not* be included as an organic compound).

In inorganic chemistry, hydrides can also serve as bridging ligands that link two metal centers in a coordination complex. This function is particularly common in group 13 elements, especially in boranes (boron hydrides) and aluminum complexes, as well as in clustered carboranes.

Hydrides

Compounds of hydrogen are often called hydrides, a term that is used fairly loosely. To chemists, the term "hydride" usually implies that the H atom has acquired a negative or anionic character, denoted H^-. The existence of the hydride anion, suggested by G.N. Lewis in 1916 for group I and II salt-like hydrides, was demonstrated by Moers in 1920 with the electrolysis of molten lithium hydride (LiH), that produced a stoichiometric quantity of hydrogen at the anode. For hydrides other than group I and II metals, the term is quite misleading, considering the low electronegativity of hydrogen. An exception in group II hydrides is BeH_2, which is polymeric. In lithium aluminum hydride, the AlH_4^- anion carries hydridic centers firmly attached to the Al(III). Although hydrides can be formed with almost all main-group elements, the number and combination of possible compounds varies widely; for example, there are over 100 binary borane hydrides known, but only one binary aluminum hydride.Binary indium hydride has not yet been identified, although larger complexes exist.

"Protons" and Acids

Oxidation of H_2 formally gives the proton, H^+. This species is central to discussion of acids, though the term proton is used loosely to refer to positively charged or cationic hydrogen, denoted H^+. A bare proton H^+ cannot exist in solution because of its strong tendency to attach itself to atoms or molecules with electrons. To avoid the convenient fiction of the naked "solvated proton" in solution, acidic aqueous solutions are sometimes considered to contain the hydronium ion (H_3O^+) organized into clusters to form $H_9O_4^+$. Other oxonium ions are found when water is in solution with other solvents.

Although exotic on earth, one of the most common ions in the universe is the H_3^+ ion, known as protonated molecular hydrogen or the triatomic hydrogen cation.

Production

H_2 is produced in chemistry and biology laboratories, often as a by-product of other reactions; in industry for the hydrogenation of unsaturated substrates; and in nature as a means of expelling reducing equivalents in biochemical reactions.

Laboratory Syntheses

In the laboratory, H_2 is usually prepared by the reaction of acids on metals such as zinc.

$$Zn + 2 H^+ \rightarrow Zn^{2+} + H_2$$

Aluminum produces H_2 upon treatment with acids but also with base:

$$2 Al + 6 H_2O \rightarrow 2 Al(OH)_3 + 3 H_2$$

The electrolysis of water is a simple method of producing hydrogen, although the resulting hydrogen necessarily has less energy content than was required to produce it. A low voltage current is run through the water, and gaseous oxygen forms at the anode while gaseous hydrogen forms at the cathode. Typically the cathode is made from platinum or another inert metal when producing hydrogen for storage. If, however, the gas is to be burnt on site, oxygen is desirable to assist the combustion, and so both electrodes would be made from inert metals. (Iron, for instance, would oxidize, and thus decrease the amount of oxygen given off.) The theoretical maximum efficiency (electricity used vs. energetic value of hydrogen produced) is between 80 – 94%. Bellona Report on Hydrogen

$$2H_2O(aq) \rightarrow 2H_2(g) + O_2(g)$$

In 2007, it was discovered that an alloy of aluminium and gallium in pellet form added to water could be used to generate hydrogen.The process creates also creates alumina, but the expensive gallium, which prevents to formation of an oxide skin on the pellets, can be re-used. This potentially has important implications for a hydrogen economy, since hydrogen can be produced on-site and does not need to be transported.

Industrial Syntheses

Hydrogen can be prepared in several different ways but the economically most important processes involve removal of hydrogen from hydrocarbons. Commercial bulk hydrogen is usually produced by the steam reforming of natural gas.At high temperatures (700 – 1100 °C; 1,300 – 2,000 °F), steam (water vapor) reacts with methane to yield carbon monoxide and H_2.

$$CH_4 + H_2O \rightarrow CO + 3\,H_2$$

This reaction is favored at low pressures but is nonetheless conducted at high pressures (20 atm; 600 inHg) since high pressure H_2 is the most marketable product. The product mixture is known as "synthesis gas" because it is often used directly for the production of methanol and related compounds. Hydrocarbons other than methane can be used to produce synthesis gas with varying product ratios. One of the many complications to this highly optimized technology is the formation of coke or carbon:

$$CH_4 \rightarrow C + 2\,H_2$$

Consequently, steam reforming typically employs an excess of H_2O.

Additional hydrogen from steam reforming can be recovered from the carbon monoxide through the water gas shift reaction, especially with an iron oxide catalyst. This reaction is also a common industrial source of carbon dioxide:

$$CO + H_2O \rightarrow CO_2 + H_2$$

Other important methods for H_2 production include partial oxidation of hydrocarbons:

$$CH_4 + 0.5\,O_2 \rightarrow CO + 2\,H_2$$

and the coal reaction, which can serve as a prelude to the shift reaction above:

$$C + H_2O \rightarrow CO + H_2$$

Hydrogen is sometimes produced and consumed in the same industrial process, without being separated. In the Haber process for the production of ammonia (the world's fifth most produced industrial compound), hydrogen is generated from natural gas.

Hydrogen is also produced in usable quantities as a co-product of the major petrochemical processes of steam cracking and reforming. Electrolysis of brine to yield chlorine also produces hydrogen as a co-product.

Biological Syntheses

H_2 is a product of some types of anaerobic metabolism and is produced by several microorganisms, usually via reactions catalyzed by iron- or nickel-containing enzymes called hydrogenases. These enzymes catalyze the reversible redox reaction between H_2 and its component two protons and two electrons. Evolution of hydrogen gas occurs in the transfer of reducing equivalents produced during pyruvate fermentation to water.

Water splitting, in which water is decomposed into its component protons, electrons, and oxygen, occurs in the light reactions in all photosynthetic organisms. Some such organisms — including the alga *Chlamydomonas reinhardtii* and cyanobacteria – have evolved a second step in the dark reactions in which protons and electrons are reduced to form H_2 gas by specialized hydrogenases in the chloroplast.Efforts have been undertaken to genetically modify cyanobacterial hydrogenases to efficiently synthesize H_2 gas even in the presence of oxygen.

Other rarer but mechanistically interesting routes to H_2 production also exist in nature. Nitrogenase produces approximately one equivalent of H_2 for each equivalent of N_2 reduced to ammonia. Some phosphatases reduce phosphite to H_2.

Applications

Large quantities of H_2 are needed in the petroleum and chemical industries. The largest application of H_2 is for the processing ("upgrading") of fossil fuels, and in the production of ammonia. The key consumers of H_2 in the petrochemical plant include hydrodealkylation, hydrodesulfurization, and hydrocracking.H_2 has several other important uses. H_2 is used as a hydrogenating agent, particularly in increasing the level of saturation of unsaturated fats and oils (found in items such as margarine), and in the production of methanol. It is similarly the source of hydrogen in the manufacture of hydrochloric acid. H_2 is also used as a reducing agent of metallic ores.

Apart from its use as a reactant, H_2 has wide applications in physics and engineering. It is used as a shielding gas in welding methods such as atomic hydrogen welding. H_2 is used as the rotor coolant in electrical generators at power stations, because it has the highest thermal conductivity of any gas. Liquid H_2 is used in cryogenic research, including superconductivity studies. Since H_2 is lighter than air, having a little more than 1/15th of the density of air, it was once widely used as a lifting agent in balloons and airships. However, this use was curtailed after the Hindenburg disaster convinced the public that the gas was too dangerous for this purpose. Hydrogen is still regularly used for the inflation of weather balloons.

Hydrogen's rarer isotopes also each have specific applications. Deuterium (hydrogen-2) is used in nuclear fission applications as a moderator to slow neutrons, and in nuclear fusion reactions. Deuterium compounds have applications in chemistry and biology in studies of reaction isotope effects. Tritium (hydrogen-3), produced in nuclear reactors, is used in the production of hydrogen bombs, as an isotopic label in the biosciences, and as a radiation source in luminous paints.

The triple point temperature of equilibrium hydrogen is a defining fixed point on the ITS-90 temperature scale.

Hydrogen as an Energy Carrier

Hydrogen is not an energy source, except in the hypothetical context of commercial nuclear fusion power plants using deuterium or tritium, a technology presently far from development. The sun's energy comes from nuclear fusion of hydrogen but this process is difficult to achieve on earth. Elemental hydrogen from solar, biological, or electrical sources costs more in energy to make than is obtained by burning it. Hydrogen may be obtained from fossil sources (such as methane) for less energy than required to make it, but these sources are unsustainable, and are also themselves direct energy sources (and are rightly regarded as the basic source of the energy in the hydrogen obtained from them).

Molecular hydrogen has been widely discussed in the context of energy, as a possible carrier of energy on an economy-wide scale. A theoretical advantage of using H_2 as an energy carrier is the localization and concentration of environmentally unwelcome aspects of hydrogen manufacture from fossil fuel energy sources. For example, CO_2 sequestration followed by carbon capture and storage could be conducted at the point of H_2 production from methane. Hydrogen used in transportation would burn cleanly, without carbon emissions. However, the infrastructure costs associated with full conversion to a hydrogen economy would be substantial. In addition, the energy density of both liquid hydrogen and hydrogen gas at any practicable pressure is significantly less than that of traditional fuel sources.

GROUP 1 (ALKALI METALS)

Some atomic and physical properties of the Group 1 elements - lithium, sodium, potassium, rubidium and caesium. You will find separate sections below covering the trends in atomic radius, first ionisation energy, electronegativity, melting and boiling points, and density.The same ideas tend to recur throughout the atomic properties.

Atomic Radius

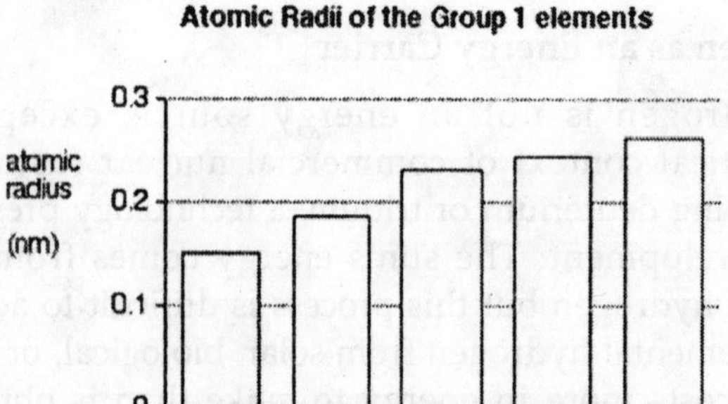

The atomic radius increases as you go down the Group.The radius of an atom is governed by

a. The number cf layers of electrons around the nucleus

b. The pull the outer electrons feel from the nucleus.

Compare lithium and sodium:

Li $1s^2 2s^1$

Na $1s^2 2s^2 2p^6 3s^1$

In each case, the outer electron feels a net pull of 1+ from the nucleus. The positive charge on the nucleus is cut down by the negativeness of the inner electrons.

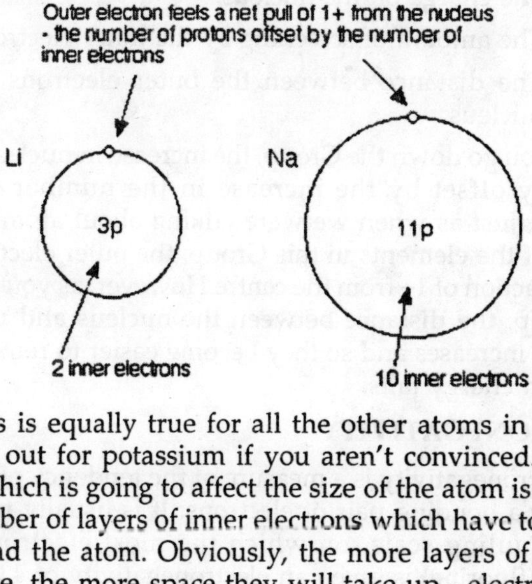

This is equally true for all the other atoms in Group 1. Work it out for potassium if you aren't convinced.The only factor which is going to affect the size of the atom is therefore the number of layers of inner electrons which have to be fitted in around the atom. Obviously, the more layers of electrons you have, the more space they will take up - electrons repel each other. That means that the atoms are bound to get bigger as you go down the Group.

First Ionisation Energy

First ionisation energy is the energy needed to remove the most loosely held electron from each of one mole of gaseous atoms to make one mole of singly charged gaseous ions - in other words, for 1 mole of this process:

$$X_{(g)} \longrightarrow X^{+}_{(g)} + e^{-}$$

First Ionisation Energy of the Group 1 elements

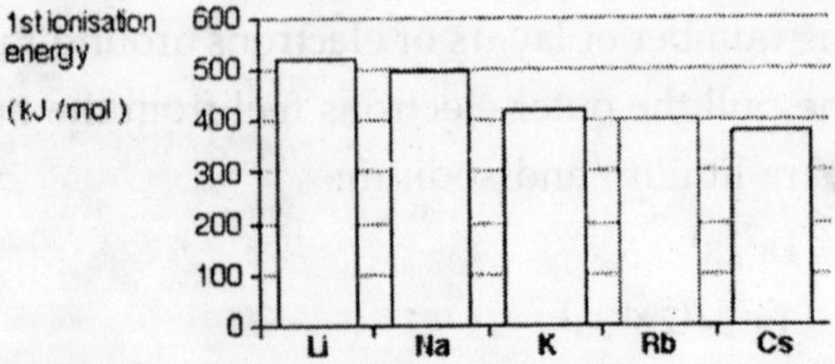

First ionisation energy falls as you go down the group.Ionisation energy is governed by

1. The charge on the nucleus,
2. The amount of screening by the inner electrons,
3. The distance between the outer electrons and the nucleus.

As you go down the Group, the increase in nuclear charge is exactly offset by the increase in the number of inner electrons. Just as when we were talking about atomic radius in each of the elements in this Group, the outer electrons feel a net attraction of 1+ from the centre.However, as you go down the Group, the distance between the nucleus and the outer electrons increases and so they become easier to remove - the ionisation energy falls.

ELECTRONEGATIVITY

Electronegativity is a measure of the tendency of an atom to attract a bonding pair of electrons. It is usually measured on the Pauling scale, on which the most electronegative element (fluorine) is given an electronegativity of 4.0.

Electronegativities of the Group 1 elements

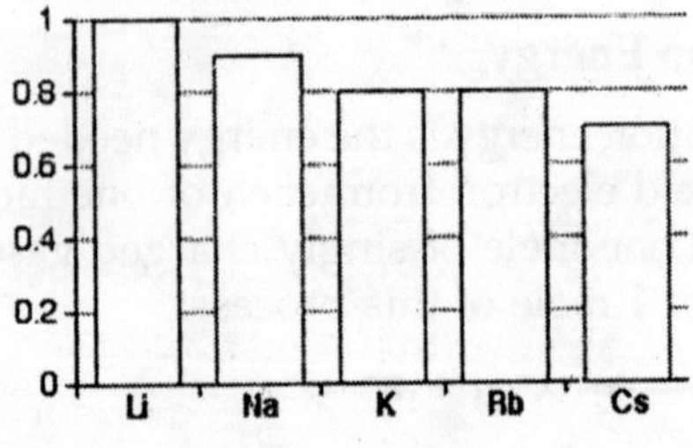

All of these elements have a very low electronegativity.Remember that the most electronegative element, fluorine, has an electronegativity of 4.0.Notice that electronegativity falls as you go down the Group. The atoms become less and less good at attracting bonding pairs of electrons.Imagine a bond between a sodium atom and a chlorine atom. Think of it to start with as a covalent bond - a pair of shared electrons. The electron pair will be dragged towards the chlorine because there is a much greater net pull from the chlorine nucleus than from the sodium one.

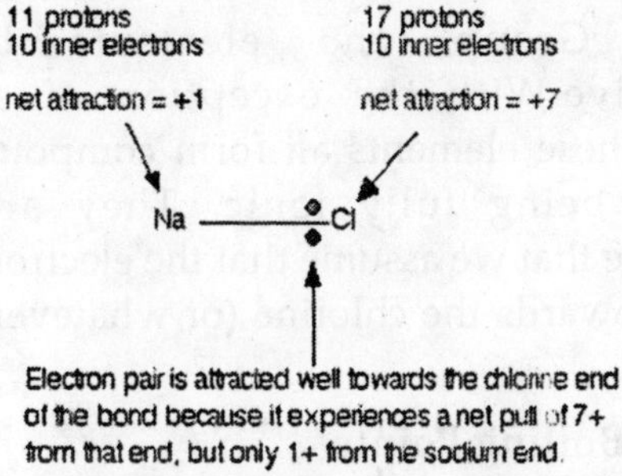

The electron pair ends up so close to the chlorine that there is essentially a transfer of an electron to the chlorine - ions are formed.The large pull from the chlorine nucleus is why chlorine is much more electronegative than sodium is.Now compare this with the lithium-chlorine bond.The net pull from each end of the bond is the same as before, but you have to remember that the lithium atom is smaller than a sodium atom. That means that the electron pair is going to be closer to the net 1+ charge from the lithium end, and so more strongly attracted to it.

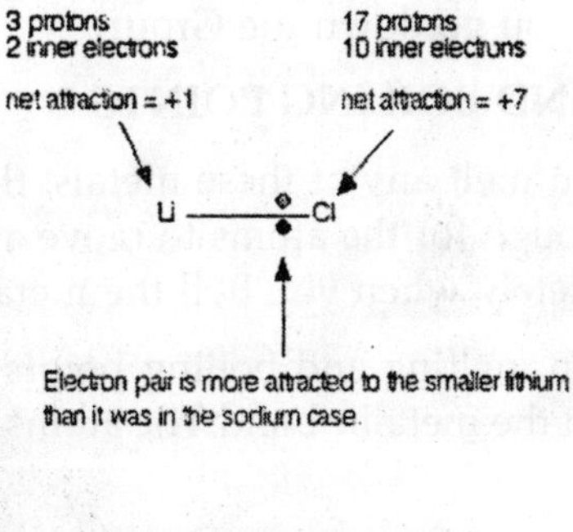

In some lithium compounds there is often a degree of covalent bonding that isn't there in the rest of the Group. Lithium iodide, for example, will dissolve in organic solvents - a typical property of covalent compounds. The iodine atom is so large that the pull from the iodine nucleus on the pair of electrons is relatively weak, and so a fully ionic bond isn't formed.

As the metal atoms get bigger, any bonding pair gets further and further away from the metal nucleus, and so is less strongly attracted towards it. In other words, as you go down the Group, the elements become less electronegative.With the exception of some lithium compounds, these elements all form compounds which we consider as being fully ionic. They are so weakly electronegative that we assume that the electron pair is pulled so far away towards the chlorine (or whatever) that ions are formed.

Melting and Boiling Points

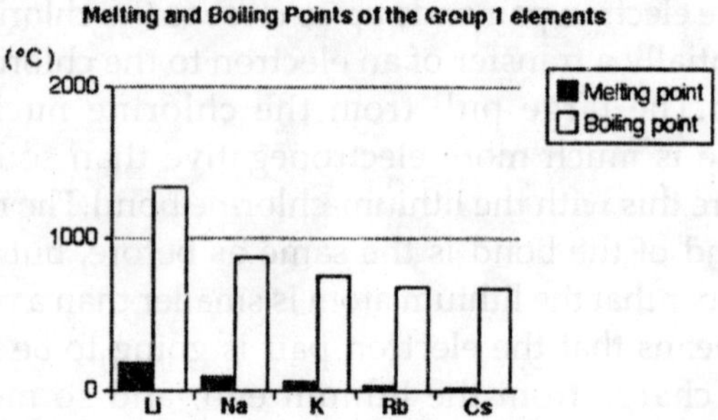

You will see that both the melting points and boiling points fall as you go down the Group.

MELTING AND BOILING POINTS

When you melt any of these metals, the metallic bond is weakened enough for the atoms to move around, and is then broken completely when you boil the metal.

The fall in melting and boiling points reflects the fall in the strength of the metallic bond.The atoms in a metal are held together by the attraction of the nuclei to the delocalised electrons. As the atoms get bigger, the nuclei get further away from these delocalised electrons, and so the attractions fall. That means that the atoms are more easily pulled apart to make a liquid and finally a gas.In the same way that we have already discussed, each of these atoms has a net pull from the nuclei of 1+. The increased charge on the nucleus as you go down the Group is offset by additional levels of screening electrons. All that matters is the distance between the nucleus and the bonding electrons.

Density

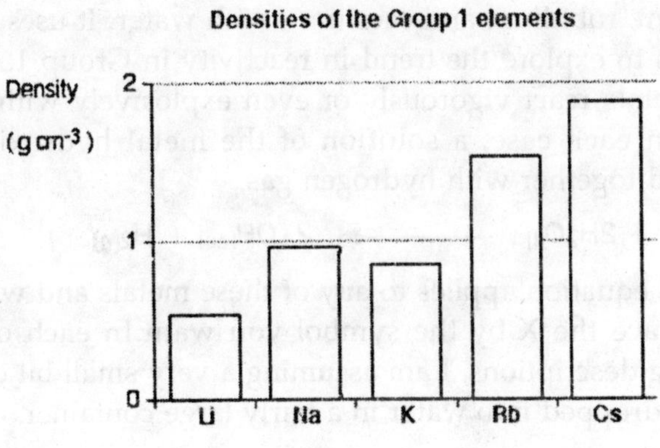

Notice that these are all light metals - and that the first three in the Group are less dense than water (less than 1 g cm^{-3}). That means that the first three will float on water, while the other two sink.The density tends to increase as you go down the Group (apart from the fluctuation at potassium).It is quite difficult to come up with a simple explanation for this, because the density depends on two factors, both of which are changing as you go down the Group.All of these metals have their atoms packed in the same way, so all you have to consider is how many atoms you can pack in a given volume, and what the mass of the individual atoms is. How many you can pack depends, of course, on their volume - and their volume, in turn, depends on their atomic radius.As you go down the Group, the atomic radius increases, and so the volume of the atoms increases as well. That means that you can't pack as many sodium atoms into a given volume as you can lithium atoms.However, as you go down the Group, the mass of the atoms increases. That means that a particular number of sodium atoms will weigh more than the same number of lithium atoms.So 1 cm^3 of sodium will contain fewer atoms than the same volume of lithium, but each atom will weigh more.

REACTIONS OF THE GROUP 1 ELEMENTS WITH WATER

The reactions of the Group 1 elements - lithium, sodium, potassium, rubidium and caesium - with water. It uses these reactions to explore the trend in reactivity in Group 1.All of these metals react vigorously or even explosively with cold water. In each case, a solution of the metal hydroxide is produced together with hydrogen gas.

$$2X_{(s)} + 2H_2O_{(l)} \longrightarrow 2XOH_{(aq)} + H_{2(g)}$$

This equation applies to any of these metals and water - just replace the X by the symbol you want.In each of the following descriptions, I am assuming a very small bit of the metal is dropped into water in a fairly large container.

Lithium

Lithium's density is only about half that of water so it floats on the surface, gently fizzing and giving off hydrogen. It gradually reacts and disappears, forming a colourless solution of lithium hydroxide. The reaction generates heat too slowly and lithium's melting point is too high for it to melt.

Sodium

Sodium also floats on the surface, but enough heat is given off to melt the sodium (sodium has a lower melting point than lithium and the reaction produces heat faster) and it melts almost at once to form a small silvery ball that dashes around the surface. A white trail of sodium hydroxide is seen in the water under the sodium, but this soon dissolves to give a colourless solution of sodium hydroxide.The sodium moves because it is pushed around by the hydrogen which is given off during the reaction. If the sodium becomes trapped on the side of the container, the hydrogen may catch fire to burn with an orange flame. The colour is due to contamination of the normally blue hydrogen flame with sodium compounds.

Potassium

Potassium behaves rather like sodium except that the reaction is faster and enough heat is given off to set light to the hydrogen. This time the normal hydrogen flame is contaminated by potassium compounds and so is coloured lilac (a faintly bluish pink).

Rubidium

Rubidium is denser than water and so sinks. It reacts violently and immediately, with everything spitting out of the container again. Rubidium hydroxide solution and hydrogen are formed.

Caesium

Caesium explodes on contact with water, quite possibly shattering the container. Caesium hydroxide and hydrogen are formed

The Group 1 metals become more reactive towards water as you go down the Group.

The overall enthalpy changes

The table gives estimates of the enthalpy change for each of the elements undergoing the reaction:

$$X_{(s)} + H_2O_{(l)} \longrightarrow XOH_{(aq)} + \frac{1}{2}H_{2(g)}$$

enthalpy change (kJ / mol)

Li	-222
Na	-184
K	-196
Rb	-195
Cs	-203

When these reactions happen, the differences between them lie entirely in what is happening to the metal atoms present. In each case, you start with metal atoms in a solid and end up with metal ions in solution.Overall, what happens to the metal is this:

$$X_{(s)} \longrightarrow X^+_{(aq)} + e^-$$

You can calculate the overall enthalpy change for this process by using Hess's Law and breaking it up into several steps that we know the enthalpy changes for.First, you would need to supply *atomisation energy* to give gaseous atoms of the metal.

$$X_{(s)} \longrightarrow X_{(g)}$$

Then ionise the metal by supplying its first *ionisation energy*.

$$X_{(g)} \longrightarrow X^+_{(g)} + e^-$$

And finally, you would get *hydration enthalpy* released when the gaseous ion comes into contact with water.

$$X^+_{(g)} + (aq) \longrightarrow X^+_{(aq)}$$

If we put values for all these steps into a table, they look like this (all values in kJ / mol):

	at. Energy	1st IE	hydr. Enthalpy	total
Li	+161	+519	-519	+161
Na	+109	+494	-406	+197
K	+90	+418	-322	+186
Rb	+86	+402	-301	+187
Cs	+79	+376	-276	+179

The atomisation energy is a measure of the strength of the metallic bond in each element. This is falling as the atom gets bigger and the metallic bond is getting longer. The delocalized electrons are further from the attraction of the nuclei in the bigger atoms.The first ionisation energy is falling because the electron being removed is getting more distant from the nucleus.The extra protons in the nucleus are screened by additional layers of electrons.The hydration enthalpy is a measure of the attraction between the metal ions and lone pairs on water molecules. As the ions get bigger, the water molecules are further from the attraction of the nucleus. The extra protons in the nucleus are again screened by the extra layers of electrons.What is happening is that the various factors are falling at different rates. That destroys any overall pattern.It is, however, possible to look at the table again and find a pattern which is useful.Let's take the last table and just look at the energy input terms - the two processes where you have to supply energy to make them work. In other words, we will miss out the hydration enthalpy term and just add up the other two.

	at. energy	1st IE	total
Li	+161	+519	+680
Na	+109	+494	+603
K	+90	+418	+508
Rb	+86	+402	+488
Cs	+79	+376	+455

There is a steady fall as you go down the Group. As you go from lithium to caesium, you need to put less energy into the reaction to get a positive ion formed. This energy will be recovered later on, but has to be supplied initially. This is going to be related to the activation energy of the reaction.The lower the activation energy, the faster the reaction.So although lithium releases most heat during the reaction, it does it relatively slowly - it isn't all released in one short, sharp burst. Caesium, on the other hand, has a significantly lower activation energy, and so although it doesn't release quite as much heat overall, it does it extremely quickly - and you get an explosion.The reactions become easier as the energy needed to form positive ions falls. This is in part due to a decrease in ionisation energy as you go down the Group, and in part to a fall in atomisation energy reflecting weaker metallic bonds as you go from lithium to caesium. This leads to lower activation energies, and therefore faster reactions.

REACTIONS WITH AIR OR OXYGEN

These are all very reactive metals and have to be stored out of contact with air to prevent their oxidation. Reactivity increases as you go down the Group.Lithium, sodium and potassium are stored in oil. (Lithium in fact floats on the oil, but there will be enough oil coating it to give it some protection. It is, anyway, less reactive than the rest of the Group.)Rubidium and caesium are normally stored in sealed glass tubes to prevent air getting at them. They are stored either in a vacuum or in an inert atmosphere of, say, argon. The tubes are broken open when the metal is used.Depending on how far down the Group you are, different kinds of oxide are formed when the metals burn. Reaction with oxygen is just a more dramatic version of the reaction with air.Lithium is unique in the Group because it also reacts with the nitrogen in the air to form lithium nitride.

Lithium

Lithium burns with a strongly red-tinged flame if heated in air. It reacts with oxygen in the air to give white lithium oxide. With pure oxygen, the flame would simply be more intense.

$$4Li + O_2 \longrightarrow 2Li_2O$$

For the record, it also reacts with the nitrogen in the air to give lithium nitride. Lithium is the only element in this Group to form a n'tride in this way.

$$6Li + N_2 \longrightarrow 2Li_3N$$

Sodium

Small pieces of sodium burn in air with often little more than an orange glow. Using larger amounts of sodium or burning it in oxygen gives a strong orange flame. You get a white solid mixture of sodium oxide and sodium peroxide.The equation for the formation of the simple oxide is just like the lithium one.

$$4Na + O_2 \longrightarrow 2Na_2O$$

The peroxide equation is:

$$2Na + O_2 \longrightarrow Na_2O_2$$

Potassium

Small pieces of potassium heated in air tend to just melt and turn instantly into a mixture of potassium peroxide and potassium superoxide without any flame being seen. Larger pieces of potassium burn with a lilac flame.The equation for the formation of the peroxide is just like the sodium one above:

$$2K + O_2 \longrightarrow K_2O_2$$

. . . and for the superoxide:

$$K + O_2 \longrightarrow KO_2$$

Rubidium and caesium

Both metals catch fire in air and produce superoxides, RbO_2 and CsO_2. The equations are the same as the equivalent potassium one.Both superoxides are described in most sources as being either orange or yellow.

Lithium (and to some extent sodium) form simple oxides, X_2O, which contain the common O^{2-} ion.

Sodium (and to some extent potassium) form peroxides, X_2O_2, containing the more complicated O_2^{2-} ion.

Potassium, rubidium and caesium form superoxides, XO_2. The structure of the superoxide ion, O_2^-, is too difficult to discuss at this level, needing a good knowledge of molecular orbital theory to make sense of it.

The more complicated ions aren't stable in the presence of a small positive ion. Consider the peroxide ion, for example.

The peroxide ion, O_2^{2-} looks like this:

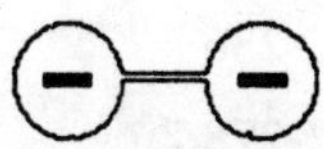

Two oxygen atoms each carrying a negative charge, and joined by a covalent bond.

The covalent bond between the two oxygen atoms is relatively weak.

Now imagine bringing a small positive ion close to the peroxide ion. Electrons in the peroxide ion will be strongly attracted towards the positive ion. This is then well on the way to forming a simple oxide ion if the right-hand oxygen atom breaks off.

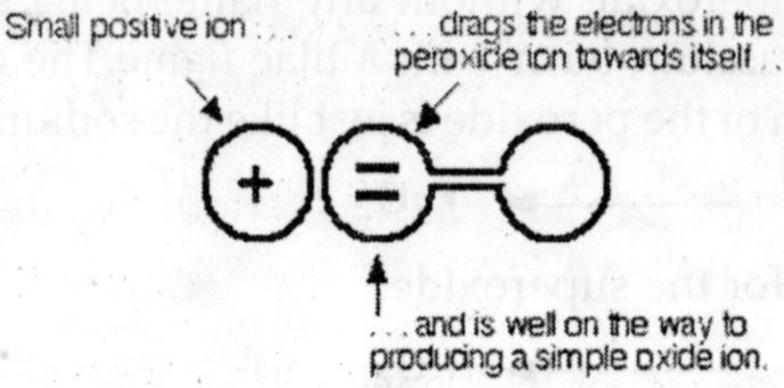

We say that the positive ion *polarises* the negative ion. This works best if the positive ion is small and highly charged - if it has a *high charge density*.Even though it only has one charge, the lithium ion at the top of the Group is so small and has such a high charge density that any peroxide ion near it falls to pieces to give an oxide and oxygen. As you go down the Group to sodium and potassium the positive ions get bigger and they don't have so much effect on the peroxide ion.The superoxide ions are even more easily pulled apart, and these are only stable in the presence of the big ions towards the bottom of the Group.So why do any of the metals form the more complicated oxides? It is a matter of energetics.In the presence of sufficient oxygen, they produce the compound whose formation gives out most energy. That gives the most stable compound.The amount of heat evolved per mole of rubidium in forming its various oxides is:

	enthalpy change (kJ / mol of Rb)
Rb_2O	-169.5
Rb_2O_2	-236
RbO_2	-278.7

The values for the various potassium oxides show exactly the same trends. As long as you have enough oxygen, forming the peroxide releases more energy per mole of metal than forming the simple oxide. Forming the superoxide releases even more.Forming the more complicated oxides from the metals releases more energy and makes the system more energetically stable. BUT . . . this only works for the metals in the lower half of the Group where the metal ions are big and have a low charge density.At the top of the Group, the small ions with a higher charge density tend to polarise the more complicated oxide ions to the point of destruction.

Reaction with Water

These are simple basic oxides, reacting with water to give the metal hydroxide.For example, lithium oxide reacts with water to give a colourless solution of lithium hydroxide.

$$X_2O + H_2O \longrightarrow 2XOH$$

Reaction with Dilute Acids

These simple oxides all react with an acid to give a salt and water. For example, sodium oxide will react with dilute hydrochloric acid to give colourless sodium chloride solution and water.

$$X_2O + 2HCl \longrightarrow 2XCl + H_2O$$

Reaction with water

If the reaction is done ice cold (and the temperature controlled so that it doesn't rise even though these reactions are strongly exothermic), a solution of the metal hydroxide and hydrogen peroxide is formed.

$$X_2O_2 + 2H_2O \longrightarrow 2XOH + H_2O_2$$

If the temperature increases (as it inevitably will unless the peroxide is added to water very, very, very slowly!), the hydrogen peroxide produced decomposes into water and oxygen. The reaction can be very violent overall.

Reaction with Dilute Acids

These reactions are even more exothermic than the ones with water. A solution containing a salt and hydrogen peroxide is formed. The hydrogen peroxide will decompose to give water and oxygen if the temperature rises - again, it is almost impossible to avoid this. Another potentially violent reaction!

$$X_2O_2 + 2HCl \longrightarrow 2XCl + H_2O_2$$

Reaction with Water

This time, a solution of the metal hydroxide and hydrogen peroxide is formed, but oxygen gas is given off as well. Once again, these are strongly exothermic reactions and the heat produced will inevitably decompose the hydrogen peroxide to water and more oxygen. Again violent!

$$2XO_2 + 2H_2O \longrightarrow 2XOH + H_2O_2 + O_2$$

Reaction with Dilute Acids

Again, these reactions are even more exothermic than the ones with water. A solution containing a salt and hydrogen peroxide is formed together with oxygen gas. The hydrogen peroxide will again decompose to give water and oxygen as the temperature rises. Violent!

$$2XO_2 + 2HCl \longrightarrow 2XCl + H_2O_2 + O_2$$

The Reactions of the Elements with Chlorine

This is included because of the similarity in appearance between the reactions of the Group 1 metals with chlorine and with oxygen.Sodium, for example, burns with an intense orange flame in chlorine in exactly the same way that it does in pure oxygen. The rest also behave the same in both gases.In each case, there is a white solid residue which is the simple chloride, XCl. There is nothing in any way complicated about these reactions!

$$2X + Cl_2 \longrightarrow 2XCl$$

Some compounds of the Group 1 elements (lithium, sodium, potassium, rubidium and caesium). You will find some information about the nitrates, carbonates, hydrogencarbonates and hydrides of the metals.We will first look at what happens to some of the compounds on heating, and then their solubility.

Group 1 compounds are more stable to heat than the corresponding compounds in Group 2. That the lithium compounds behave similarly to Group 2 compounds, but the rest of Group 1 are in some way different.

Heating the Nitrates

Most nitrates tend to decompose on heating to give the metal oxide, brown fumes of nitrogen dioxide, and oxygen.

For example, a typical *Group 2 nitrate* like magnesium nitrate decomposes like this:

$$2Mg(NO_3)_{2(s)} \longrightarrow 2MgO_{(s)} + 4NO_{2(g)} + O_{2(g)}$$

In *Group 1,* lithium nitrate behaves in the same way - producing lithium oxide, nitrogen dioxide and oxygen.

$$4LiNO_{3(s)} \longrightarrow 2Li_2O_{(s)} + 4NO_{2(g)} + O_{2(g)}$$

The rest of the Group, however, don't decompose so completely (at least not at Bunsen temperatures) - producing the metal nitrite and oxygen, but no nitrogen dioxide.

$$2XNO_{3(s)} \longrightarrow 2XNO_{2(s)} + O_{2(g)}$$

All the nitrates from sodium to caesium decomposes in this same way, the only difference being how hot they have to be to undergo the reaction. As you go down the Group, the decomposition gets more difficult, and you have to use higher temperatures.

Heating the Carbonates

Most carbonates tend to decompose on heating to give the metal oxide and carbon dioxde.For example, a typical *Group 2 carbonate* like calcium carbonate decomposes like this:

$$CaCO_{3(s)} \longrightarrow CaO_{(s)} + CO_{2(g)}$$

In *Group 1,* lithium carbonate behaves in the same way - producing lithium oxide and carbon dioxide.

$$Li_2CO_{3(s)} \longrightarrow Li_2O_{(s)} + CO_{2(g)}$$

The rest of the Group 1 carbonates don't decompose at Bunsen temperatures, although at higher temperatures they will. The decomposition temperatures again increase as you go down the Group.

The Thermal Stability of the Hydrogencarbonates

The *Group 2* hydrogencarbonates like calcium hydrogencarbonate are so unstable to heat that they only exist in solution. Any attempt to get them out of solution causes them to decompose to give the carbonate, carbon dioxide and water.

$$Ca(HCO_3)_{2(aq)} \longrightarrow CaCO_{3(s)} + CO_{2(g)} + H_2O_{(l)}.$$

By contrast, the *Group 1* hydrogencarbonates are stable enough to exist as solids, although they do decompose easily on heating. For example, for sodium hydrogencarbonate:

$$2NaHCO_{3(s)} \longrightarrow Na_2CO_{3(s)} + CO_{2(g)} + H_2O_{(l)}$$

Detailed explanations are given for the carbonates because the diagrams are easier to draw. Exactly the same arguments apply to the nitrates or hydrogencarbonates.There are two ways of explaining the increase in thermal stability as you go down the Group. The hard way is in terms of the energetics of the process; the simple way is to look at the polarising ability of the positive ions.A small positive ion has a lot of charge packed into a small volume of space - especially if it has more than one positive charge. It has a high *charge density* and will have a marked distorting effect on any negative ions which happen to be near it.A bigger positive ion has the same charge spreaa over a larger volume of space. Its charge density will be lower, and it will cause less distortion to nearby negative ions.

Carbonate ion

If you worked out the structure of a carbonate ion using "dots-and-crosses" or some similar method, you would probably come up with:

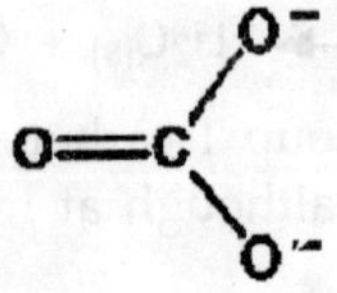

This shows two single carbon-oxygen bonds and one double one, with two of the oxygens each carrying a negative charge. Unfortunately, in real carbonate ions all the bonds are identical, and the charges are spread out over the whole ion - although concentrated on the oxygen atoms. We say that the charges are *delocalised*.This is a rather more complicated version of the bonding you might have come across in benzene or in ions like ethanoate. For the purposes of this topic, you don't need to understand how this bonding has come about.The next diagram shows the delocalised electrons. The shading is intended to show that there is a greater chance of finding them around the oxygen atoms than near the carbon.

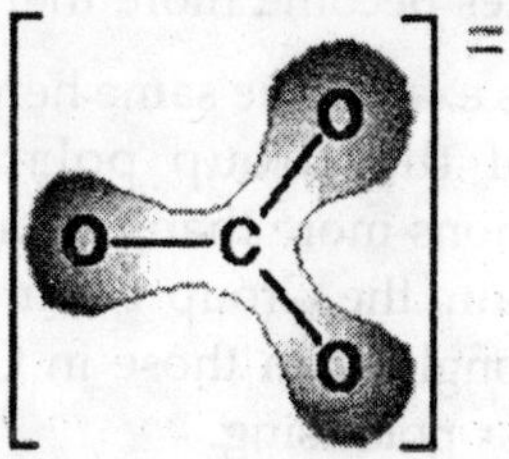

Polarising the Carbonate Ion

Now imagine what happens when this ion is placed next to a positive ion. The positive ion attracts the delocalised electrons in the carbonate ion towards itself. The carbonate ion becomes polarised. The diagram shows what happens with an ion from Group 2, carrying two positive charges

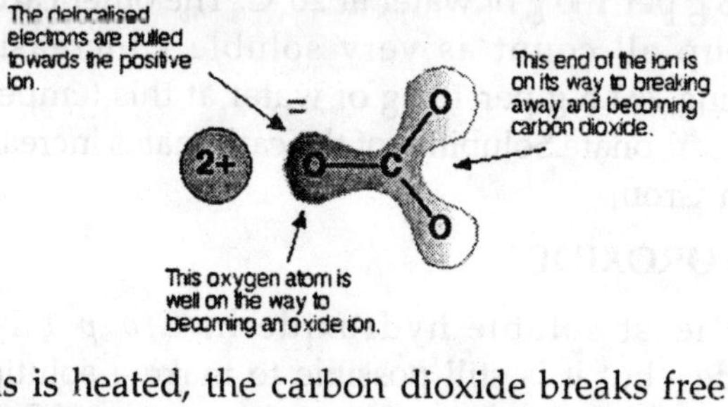

If this is heated, the carbon dioxide breaks free to leave the metal oxide.How much you need to heat the carbonate before that happens depends on how polarised the ion was. If it is highly polarised, you need less heat than if it is only

slightly polarised.If the positive ion only had one positive charge, the polarising effect would be less. That is why the Group 1 compounds are more thermally stable than those in Group 2. You have to heat the Group 1 compound more because the carbonate ions are less polarised by singly charged positive ions. The smaller the positive ion is, the higher the charge density, and the greater effect it will have on the carbonate ion. As the positive ions get bigger as you go down the Group, they have less effect on the carbonate ions near them. To compensate for that, you have to heat the compound more in order to persuade the carbon dioxide to break free and leave the metal oxide.In other words, as you go down the Group, the carbonates become more thermally stable.

The argument is exactly the same here. The small positive ions at the top of the Group polarise the nitrate or hydrogencarbonate ions more than the larger positive ions at the bottom.And, again, the Group 1 compounds will need to be heated more strongly than those in Group 2 because the Group 1 ions are less polarising.

The Carbonates

For example, *Group 2* carbonates are virtually insoluble in water. Magnesium carbonate is soluble to the extent of about 0.02 g per 100 g of water at room temperature.By contrast, the *least* soluble *Group 1* carbonate is lithium carbonate. A saturated solution of it has a concentration of about 1.3 g per 100 g of water at 20°C. The other carbonates in the Group all count as very soluble - increasing to an astonishing 261.5 g per 100 g of water at this temperature for caesium carbonate.Solubility of the carbonates increases as you go down Group 1.

THE HYDROXIDES

The least soluble hydroxide in *Group 1* is lithium hydroxide - but it is still possible to make a solution with a concentration of 12.8 g per 100 g of water at 20°C. The other hydroxides in the Group are even more soluble.Solubility of the hydroxides increases as you go down Group 1.In *Group 2*, the *most* soluble one is barium hydroxide - and it is only possible to make a solution of concentration around 3.9 g per 100 g of water at the same temperature.Explaining the trends in Group 2 was difficult enough. Comparing them with Group 1 is going to be even more difficult - particularly in the case of the carbonates, because the trends in the two Groups are in opposite directions. The carbonates get more soluble as you go down Group 1, but tend to get less soluble down Group 2.

Saline (salt-like) Hydrides

The hydrides of Group 1 metals are white crystalline solids which contain the metal ions and hydride ions, H^-. They have exactly the same crystal structure as sodium chloride - that's why they are called saline or salt-like hydrides.Because they can react violently with water or moist air, they are normally supplied as suspensions in mineral oil.

Preparation of the Group 1 hydrides

These are made by passing hydrogen gas over the heated metal. For example, for lithium hydride:

$$2Li + H_2 \longrightarrow 2LiH$$

Reactions of the Group 1 hydrides

ELECTROLYSIS

On heating, most of these hydrides decompose back into the metal and hydrogen before they melt. It is, however, possible to melt lithium hydride and to electrolyse the melt.The metal is released at the cathode as you would expect. Hydrogen is given off at the anode (the positive electrode) and this is evidence for the presence of the negative hydride ion in lithium hydride.The anode equation is:

$$2H^- \longrightarrow H_2 + 2e^-$$

The other Group 1 hydrides can be electrolysed in solution in various molten mixtures such as a mixture of lithium chloride and potassium chloride. Mixtures such as these melt at lower temperatures than the pure chlorides.

Reaction with Water

These hydrides react violently with water releasing hydrogen gas and producing the metal hydroxide.For example, sodium hydride reacts with water to produce a solution of sodium hydroxide and hydrogen gas.

$$NaH + H_2O \longrightarrow NaOH + H_2$$

FLAME TESTS

A flame test for a range of metal ions, and briefly describes how the flame colour arises.Flame tests are used to identify the presence of a relatively small number of metal ions in a compound. Not all metal ions give flame colours.For Group 1 compounds, flame tests are usually by far the easiest way of identifying which metal you have got. For other metals, there are usually other easy methods which are more reliable - but the flame test can give a useful hint as to where to look.

Carrying out a Flame Test

Clean a platinum or nichrome (a nickel-chromium alloy) wire by dipping it into concentrated hydrochloric acid and then holding it in a hot Bunsen flame. Repeat this until the wire doesn't produce any colour in the flame.When the wire is clean, moisten it again with some of the acid and then dip it into a small amount of the solid you are testing so that some sticks to the wire. Place the wire back in the flame again.If the flame colour is weak, it is often worthwhile to dip the wire back in the acid again and put it back into the flame as if you were cleaning it.

The Colours

The colours in the table are just a guide. Almost everybody sees and describes colours differently. for example, used the word "red" several times to describe colours which can be quite different from each other.

flame colour

Li red

Na strong persistent orange

K lilac (pink)

Rb red (reddish-violet)

Cs blue violet

Ca orange-red

Sr red

Ba pale green

Cu blue-green (often with white flashes)

Pb greyish-white

The Origin of Flame Colours

Flame colours are produced from the movement of the electrons in the metal ions present in the compounds.

For example, a sodium ion in an unexcited state has the structure $1s^2 2s^2 2p^6$.When you heat it, the electrons gain energy and can jump into any of the empty orbitals at higher levels - for example, into the 7s or 6p or 4d or whatever, depending on how much energy a particular electron happens to absorb from the flame.Because the electrons are now at a higher and more energetically unstable level, they tend to fall back down to where they were before - but not necessarily all in one go

An electron which had been excited from the 2p level to an orbital in the 7 level, for example, might jump back to the 2p level in one go. That would release a certain amount of energy which would be seen as light of a particular colour.

However, it might jump back in two (or more) stages. For example, first to the 5 level and then back to the 2 level.

Each of these jumps involves a specific amount of energy being released as light energy, and each corresponds to a particular colour.As a result of all these jumps, a spectrum of coloured lines will be produced. The colour you see will be a combination of all these individual colours.The exact sizes of the possible jumps in energy terms vary from one metal ion to another. That means that each different ion will have a different pattern of spectral lines, and so a different flame colour.

GROUP 2 (ALKALINE EARTH METALS)

The trends in some atomic and physical properties of the Group 2 elements - beryllium, magnesium, calcium, strontium and barium. The trends in atomic radius, first ionisation energy, electronegativity and melting point.Even if you aren't currently interested in all these things.The same ideas tend to recur throughout the atomic properties.

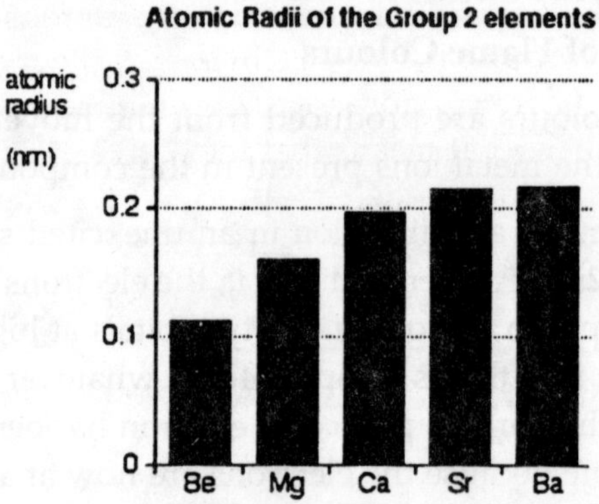

The atomic radius increases as you go down the Group. Notice that beryllium has a particularly small atom compared with the rest of the Group.The radius of an atom is governed by

1. The number of layers of electrons around the nucleus
2. The pull the outer electrons feel from the nucleus.

Compare beryllium and magnesium:

Be$1s^2 2s^2$

Mg $1s^2 2s^2 2p^6 3s^2$

In each case, the two outer electrons feel a net pull of 2+ from the nucleus. The positive charge on the nucleus is cut down by the negativeness of the inner electrons.

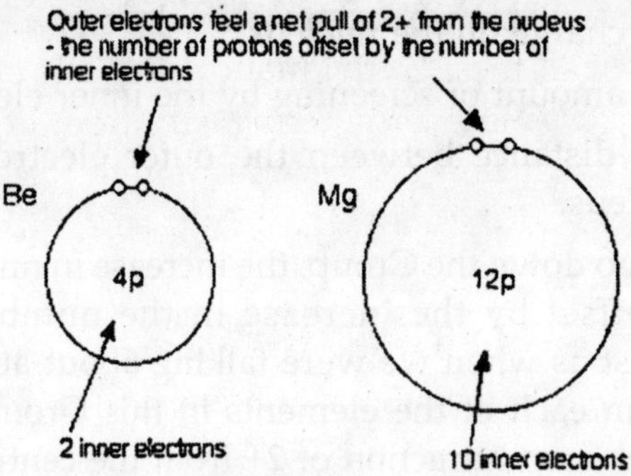

This is equally true for all the other atoms in Group 2. Work it out for calcium if you aren't convinced.

The only factor which is going to affect the size of the atom is therefore the number of layers of inner electrons which have to be fitted in around the atom. Obviously, the more layers of electrons you have, the more space they will take up - electrons repel each other. That means that the atoms are bound to get bigger as you go down the Group.

First Ionisation Energy

First ionisation energy is the energy needed to remove the most loosely held electron from each of one mole of gaseous atoms to make one mole of singly charged gaseous ions - in other words, for 1 mole of this process:

$$X_{(g)} \longrightarrow X^{+}_{(g)} + e^{-}$$

First Ionisation Energy of the Group 2 elements

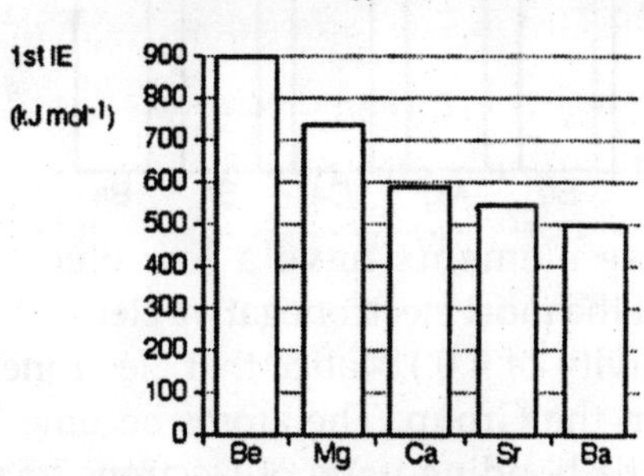

Notice that first ionisation energy falls as you go down the group.Ionisation energy is governed by

1. The charge on the nucleus,
2. The amount of screening by the inner electrons,
3. The distance between the outer electrons and the nucleus.

As you go down the Group, the increase in nuclear charge is exactly offset by the increase in the number of inner electrons. Just as when we were talking about atomic radius further up, in each of the elements in this Group, the outer electrons feel a net attraction of 2+ from the centre.However, as you go down the Group, the distance between the nucleus and the outer electrons increases and so they become easier to remove - the ionisation energy falls.

ELECTRONEGATIVITY

Electronegativity is a measure of the tendency of an atom to attract a bonding pair of electrons. It is usually measured on the Pauling scale, on which the most electronegative element (fluorine) is given an electronegativity of 4.0.

Electronegativity of the Group 2 elements

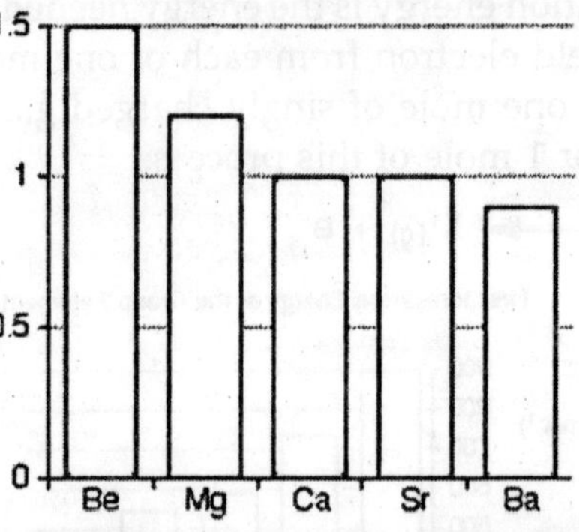

All of these elements have a low electronegativity. (Remember that the most electronegative element, fluorine, has an electronegativity of 4.0.) Notice that electronegativity falls as you go down the Group. The atoms become less and less good at attracting bonding pairs of electrons.Imagine a bond between a magnesium atom and a chlorine atom. Think of it to start with as a covalent bond - a pair of shared electrons.

The electron pair will be dragged towards the chlorine end because there is a much greater net pull from the chlorine nucleus than from the magnesium one.

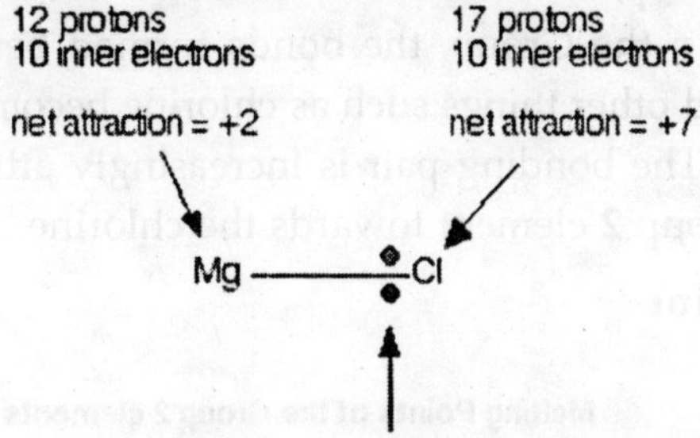

The electron pair ends up so close to the chlorine that there is essentially a transfer of an electron to the chlorine - ions are formed.The large pull from the chlorine nucleus is why chlorine is much more electronegative than magnesium is.Now compare this with the beryllium-chlorine bond.The net pull from each end of the bond is the same as before, but you have to remember that the beryllium atom is smaller than a magnesium atom. That means that the electron pair is going to be closer to the net 2+ charge from the beryllium end, and so more strongly attracted to it.

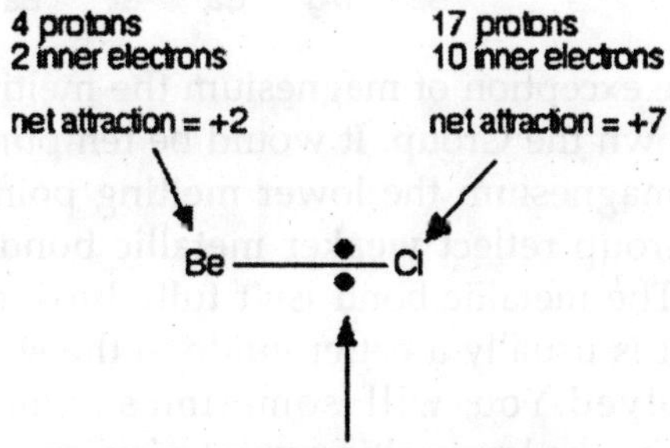

Electron pair is more attracted to the beryllium than it was in the magnesium case.

In this case, the electron pair doesn't get attracted close enough to the chlorine for an ionic bond to be formed. Because of its small size, beryllium forms covalent bonds, not ionic ones. The attraction between the beryllium nucleus and a bonding pair is always too great for ions to be formed.

As the metal atoms get bigger, any bonding pair gets further and further away from the metal nucleus, and so is less strongly attracted towards it. In other words, as you go down the Group, the elements become less electronegative.As you go down the Group, the bonds formed between these elements and other things such as chlorine become more and more ionic. The bonding pair is increasingly attracted away from the Group 2 element towards the chlorine.

Melting Point

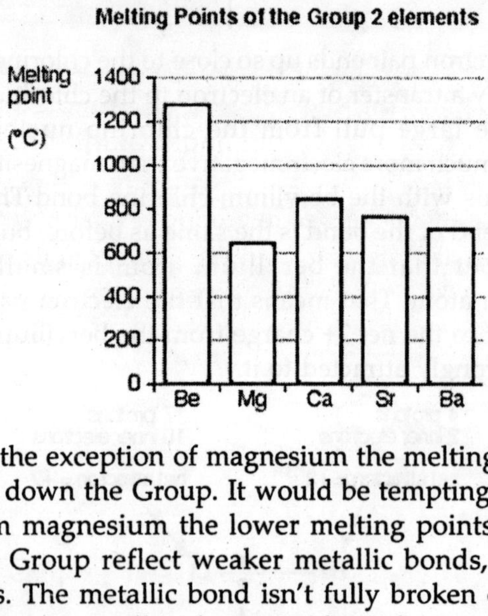

With the exception of magnesium the melting point falls as you go down the Group. It would be tempting to say that apart from magnesium the lower melting points as you go down the Group reflect weaker metallic bonds, but this is dangerous. The metallic bond isn't fully broken on melting. Boiling point is usually a better guide to the strengths of the forces involved.You will sometimes come across an explanation for the low melting point of magnesium in terms of the way the atoms pack in the crystal. It isn't very difficult to pick holes in this argument and, in any case, the boiling point of magnesium is also surprisingly low - and packing is completely irrelevant when you are thinking about a liquid. In the case of magnesium, there must be some factor producing a relatively weak metallic bond.

Boiling Points of the Group 2 elements

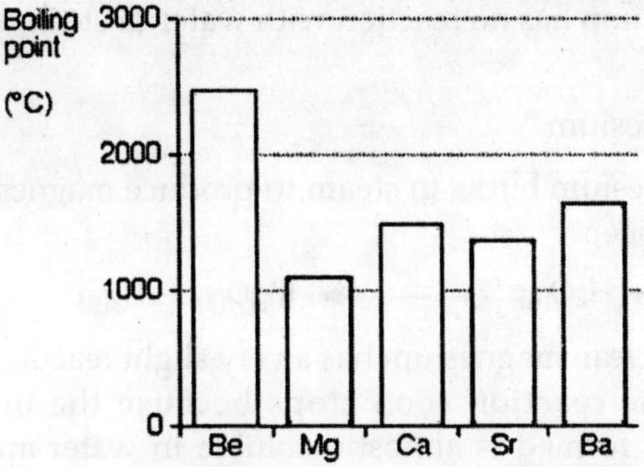

You will see that there is no obvious pattern in the boiling points - implying that there is no simple pattern in the strengths of the metallic bonds.Another measure of the strengths of the metallic bonds might be atomisation energy. This is the energy needed to produce 1 mole of separated atoms in the gas state starting from the element in its standard state.

Atomisation Energy of the Group 2 elements

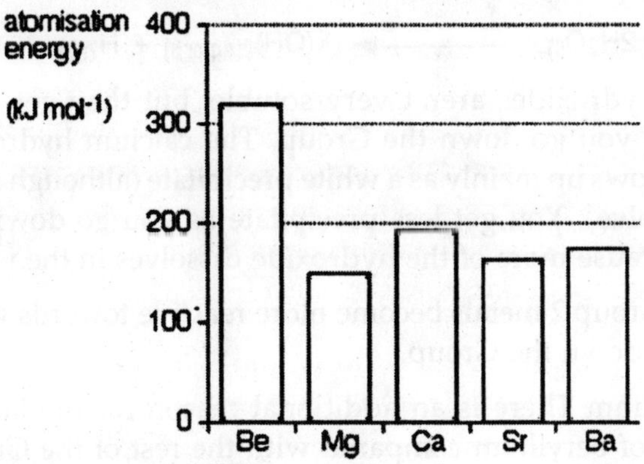

Reactions with Water

Beryllium, magnesium, calcium, strontium and barium - with water (or steam). It uses these reactions to explore the trend in reactivity in Group 2.

Beryllium

Beryllium has no reaction with water or steam even at red heat.

Magnesium

Magnesium burns in steam to produce magnesium oxide and hydrogen.

$$Mg_{(s)} + H_2O_{(g)} \longrightarrow MgO_{(s)} + H_{2(g)}$$

Very clean magnesium has a very slight reaction with cold water. The reaction soon stops because the magnesium hydroxide formed is almost insoluble in water and forms a barrier on the magnesium preventing further reaction.

$$Mg_{(s)} + 2H_2O_{(l)} \longrightarrow Mg(OH)_{2(s)} + H_{2(g)}$$

Calcium, strontium and barium

These all react with cold water with increasing vigour to give the metal hydroxide and hydrogen. Strontium and barium have reactivities similar to lithium in Group 1 of the Periodic Table.The equation for the reactions of any of these metals would be:

$$X_{(s)} + 2H_2O_{(l)} \longrightarrow X(OH)_{2(aq\ or\ s)} + H_{2(g)}$$

The hydroxides aren't very soluble, but they get more soluble as you go down the Group. The calcium hydroxide formed shows up mainly as a white precipitate (although some does dissolve). You get less precipitate as you go down the Group because more of the hydroxide dissolves in the water.

The Group 2 metals become more reactive towards water as you go down the Group.

Beryllium There is an additional reason for the lack of reactivity of beryllium compared with the rest of the Group. Beryllium has a strong resistant layer of oxide on its surface which lowers its reactivity. If you add that to the trends explained below, beryllium turns out to be very unreactive.

The enthalpy change of a reaction is a measure of the amount of heat absorbed or evolved when the reaction takes place. An enthalpy change is negative if heat is evolved, and positive if it is absorbed. If you calculate the enthalpy change for the possible reactions between beryllium or magnesium and steam, you come up with these answers:

$$Be_{(s)} + H_2O_{(g)} \longrightarrow BeO_{(s)} + H_{2(g)} \quad \Delta H = -369\ kJ\ mol^{-1}$$

$$Mg_{(s)} + H_2O_{(g)} \longrightarrow MgO_{(s)} + H_{2(g)} \quad \Delta H = -360\ kJ\ mol^{-1}$$

Notice that both possible reactions are strongly exothermic, giving out almost identical amounts of heat. However, only the magnesium reaction actually happens. The explanation for the different reactivities must lie somewhere else.Similarly, if you calculate the enthalpy changes for the reactions between calcium, strontium or barium and cold water, you again find that the amount of heat evolved in each case is almost exactly the same - in this case, about -430 kJ mol^{-1}.The reason for the increase in reactivity must again lie elsewhere.The activation energy for a reaction is the minimum amount of energy which is needed in order for the reaction to take place. It doesn't matter how exothermic the reaction would be once it got started - if there is a high activation energy barrier, the reaction will take place very slowly, if at all.When Group 2 metals react to form oxides or hydroxides, metal ions are formed.

The formation of the ions from the original metal involves various stages all of which require the input of energy - contributing to the activation energy of the reaction. These stages involve the input of:

1. The atomisation energy of the metal. This is the energy needed to break the bonds holding the atoms together in the metallic lattice.
2. The first + second ionisation energies. These are necessary to convert the metal atoms into ions with a 2+ charge.

After this, there will be a number of steps which give out heat again - leading to the formation of the products, and overall exothermic reactions.The graph shows the effect of these important energy-absorbing stages as you go down Group 2.

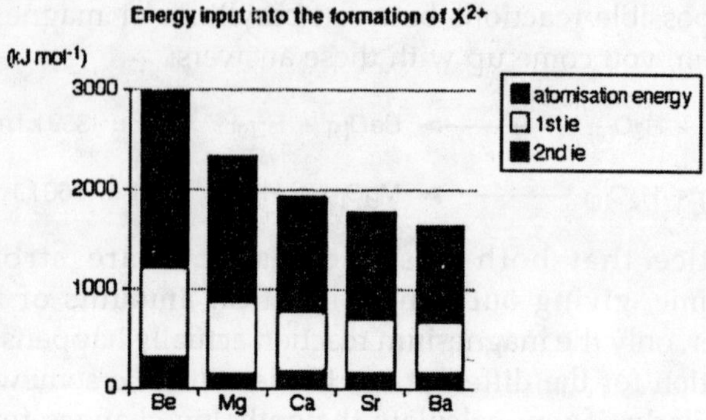

Notice that the ionisation energies dominate this - particularly the second ionisation energies. Ionisation energies fall as you go down the Group. Because it gets easier to form the ions, the reactions will happen more quickly.The reactions become easier as the energy needed to form positive ions falls. This is mainly due to a decrease in ionisation energy as you go down the Group. This leads to lower activation energies, and therefore faster reactions.

THE REACTIONS WITH OXYGEN

Formation of Simple Oxides

On the whole, the metals burn in oxygen to form a simple metal oxide.Beryllium is reluctant to burn unless it is in the form of dust or powder. Beryllium has a very strong (but very thin) layer of beryllium oxide on its surface, and this prevents any new oxygen getting at the underlying beryllium to react with it.

$$2X_{(s)} + O_{2(g)} \longrightarrow 2XO_{(s)}$$

"X" in the equation can represent any of the metals in the Group.It is almost impossible to find any trend in the way the metals react with oxygen. It would be quite untrue to say that they burn more vigorously as you go down the Group.To be able to make any sensible comparison, you would have to have pieces of metal which were all equally free of oxide coating, with exactly the same surface area and shape, exactly the same flow of oxygen around them, and heated to exactly the same extent to get them started.

Beryllium: beryllium burns with the same sort of silvery sparkles that magnesium or aluminium powder burn with if they are scattered into a flame.

Magnesium: of course, burns with a typical intense white flame.

Calcium: is quite reluctant to start burning, but then bursts dramatically into flame, burning with an intense white flame with a tinge of red at the end.

Strontium: burns with a much richer red tinge to the flame.

Barium: the flame appeared to be white with some pale green tinges. It wasn't noticeably any more dramatic than the familiar magnesium flame.

Formation of Peroxides

Strontium and barium will also react with oxygen to form strontium or barium peroxide.Strontium forms this if it is heated in oxygen under high pressures, but barium forms barium peroxide just on normal heating in oxygen. Mixtures of barium oxide and barium peroxide will be produced.

$$Ba_{(s)} + O_{2(g)} \longrightarrow BaO_{2(s)}$$

The strontium equation would look just the same.

The Reactions with Air

The reactions of the Group 2 metals with air rather than oxygen is complicated by the fact that they all react with nitrogen to produce nitrides. In each case, you will get a mixture of the metal oxide and the metal nitride.

The general equation for the Group is:

$$3X_{(s)} + N_{2(g)} \longrightarrow X_3N_{2(s)}$$

The familiar white ash you get when you burn magnesium ribbon in air is a mixture of magnesium oxide and magnesium nitride There are no simple patterns. It would be tempting to say that the reactions get more vigorous as you go down the Group, but it isn't true.The overall amount of heat evolved when one mole of oxide is produced from the metal and oxygen shows no simple pattern:

Heat evolved when 1 mole of oxide is formed from its elements

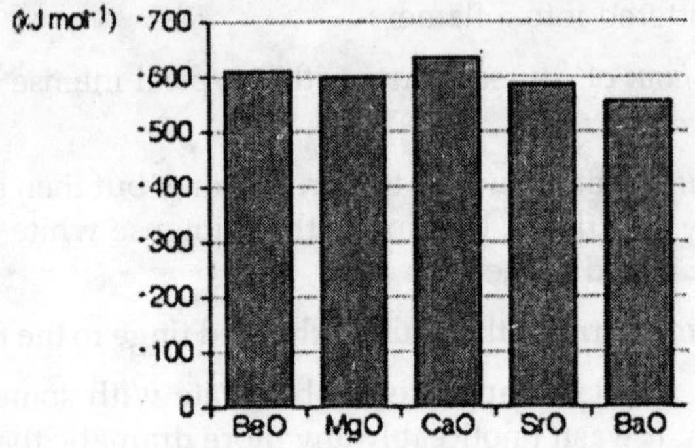

If anything, there is a slight tendency for the amount of heat evolved to get less as you go down the Group.But how reactive a metal seems to be depends on how fast the reaction happens - not the overall amount of heat evolved. The speed is controlled by factors like the presence of surface coatings on the metal and the size of the activation energy.You could argue that the activation energy will fall as you go down the Group and that will make the reaction go faster. The activation energy will fall because the ionisation energies of the metals fall.In this case, though, the effect of the fall in the activation energy is masked by other factors - for example, the presence of existing oxide layers on the metals, and the impossibility of controlling precisely how much heat you are supplying to the metal in order to get it to start burning.It is interesting to look at what happens if you heat a *very* reactive metal like potassium in air. The potassium melts at a low temperature and almost instantly turns into a pool of molten potassium oxide. The activation energy is so low that the reaction happens very quickly at quite a low temperature.Magnesium, on the other hand, has to be heated to quite a high temperature before it will start to react. The activation energy is much higher. There are also problems with surface coatings. It is then so hot that it produces the typical intense white flame.It would obviously be totally misleading to say that magnesium is more reactive than potassium on the evidence of the bright flame.

Beryllium, magnesium and calcium don't form peroxides when heated in oxygen, but strontium and barium do. There is an increase in the tendency to form the peroxide as you go down the Group.The peroxide ion, O_2^{2-} looks llike this:

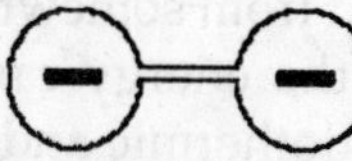

The covalent bond between the two oxygen atoms is relatively weak.Now imagine bringing a small 2+ ion close to the peroxide ion. Electrons in the peroxide ion will be strongly attracted towards the positive ion. This is then well on the way to forming a simple oxide ion if the right-hand oxygen atom breaks off.

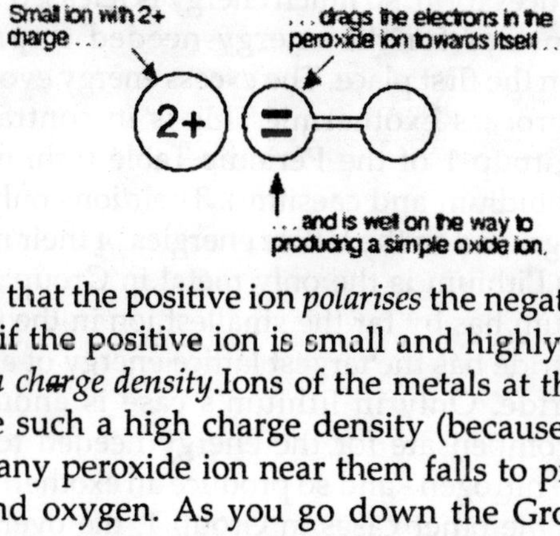

We say that the positive ion *polarises* the negative ion. This works best if the positive ion is small and highly charged - if it has a *high charge density*.Ions of the metals at the top of the Group have such a high charge density (because they are so small) that any peroxide ion near them falls to pieces to give an oxide and oxygen. As you go down the Group and the positive ions get bigger, they don't have so much effect on the peroxide ion.Barium peroxide can form because the barium ion is so large that it doesn't have such a devastating effect on the peroxide ions as the metals further up the Group.

Nitrogen is often thought of as being fairly unreactive, and yet all these metals combine with it to produce nitrides,

X_3N_2, containing X^{2+} and N^{3-} ions.Nitrogen is fairly unreactive because of the very large amount of energy needed to break the triple bond joining the two atoms in the nitrogen molecule, N_2.When something like magnesium nitride forms, you have to supply all the energy needed to form the magnesium ions as well as breaking the nitrogen-nitrogen bonds and then forming N^{3-} ions. All of these processes absorb energy.This energy has to be recovered from somewhere to give an overall exothermic reaction - if the energy can't be recovered, the overall change will be endothermic and won't happen.

Energy is evolved when the ions come together to produce the crystal lattice. This energy is known as *lattice energy* or *lattice enthalpy*.The size of the lattice energy depends on the attractions between the ions. The lattice energy is greatest if the ions are small and highly charged - the ions will be close together with very strong attractions. In the whole of Group 2, the attractions between the 2+ metal ions and the 3- nitride ions are big enough to produce very high lattice energies.When the crystal lattices form, so much energy is released that it more than compensates for the energy needed to produce the various ions in the first place. The excess energy evolved makes the overall process exothermic.This is in contrast to what happens in Group 1 of the Periodic Table (lithium, sodium, potassium, rubidium and caesium). Their ions only carry one positive charge, and so the lattice energies of their nitrides will be much less.Lithium is the only metal in Group 1 to form a nitride. Lithium has by far the smallest ion in the Group, and so lithium nitride has the largest lattice energy of any possible Group 1 nitride. Only in lithium's case is enough energy released to compensate for the energy needed to ionise the metal and the nitrogen - and so produce an exothermic reaction overall.In all the other cases in Group 1, the overall reaction would be endothermic. Those reactions don't happen, and the nitrides of sodium and the rest aren't formed.

SOLUBILITY

The solubility in water of the hydroxides, sulphates and carbonates of the Group 2 elements - beryllium, magnesium, calcium, strontium and barium.

The hydroxides become more soluble as you go down the Group.

Magnesium hydroxide appears to be insoluble in water. However, if you shake it with water, filter it and test the pH of the solution, you find that it is slightly alkaline. This shows that there are more hydroxide ions in the solution than there were in the original water. Some magnesium hydroxide must have dissolved.

Calcium hydroxide solution is used as "lime water". 1 litre of pure water will dissolve about 1 gram of calcium hydroxide at room temperature.

Barium hydroxide is soluble enough to be able to produce a solution with a concentration of around 0.1 mol dm^{-3} at room temperature.

Solubility of the Sulphates

The sulphates become less soluble as you go down the Group.The simple trend is true provided you include hydrated beryllium sulphate in it, but not if the beryllium sulphate is anhydrous.Anyhydrous beryllium sulphate, $BeSO_4$, as *insoluble* whereas the hydrated form, $BeSO_4.4H_2O$ is soluble.You are probably familiar with the reaction between magnesium and dilute sulphuric acid to give lots of hydrogen and a colourless solution of *magnesium sulphate.* Notice that you get a solution, not a precipitate. The magnesium sulphate is obviously soluble.You may also remember that *barium sulphate* is formed as a white precipitate during the test for sulphate ions in solution. The ready formation of a precipitate shows that the barium sulphate must be pretty insoluble. In fact, 1 litre of water will only dissolve about 2 mg of barium sulphate at room temperature.

Solubility of the Carbonates

The carbonates tend to become less soluble as you go down the Group.

None of the carbonates is anything more than very sparingly soluble. Magnesium carbonate is soluble. The trend to lower solubility is, however, broken at the bottom of the Group. Barium carbonate is slightly more soluble than strontium sulphate.

Enthalpy Changes

The usual explanation is in terms of the enthalpy changes which occur when an ionic compound dissolves in water. Energy has to be supplied to break up the lattice of ions, and energy is released when these ions form bonds of one sort or another with water molecules.As you go down a Group, the energy needed to break up the lattice falls as the positive ions get bigger. The bigger the ions, the more distance there is between them, and the weaker the forces holding them together.Again as the positive ions get bigger, the energy released as the ions bond to water molecules (their hydration enthalpies) falls as well. Bigger ions aren't so strongly attracted to the water molecules.Since both of these important enthalpy terms fall as you go down the Group, what matters in deciding whether the change becomes more endothermic or more exothermic overall is how fast they fall relative to each other.Here is an example of the sort of calculations you might do to work out the enthalpy change of solution for sodium chloride and potassium chloride.

	lattice enthalpy	hydration enthalpy of positive ion	hydration enthalpy of Cl^-	
NaCl	+771	-406	-364	= +1 kJ mol^{-1}
KCl	+701	-322	-364	= +15 kJ mol^{-1}
	falls by 70	falls by 84		

The enthalpy of solution changes from NaCl to KCl because the lattice enthalpy and hydration enthalpy of the positive ion fall by different amounts.

1. If the hydration enthalpy falls faster than the lattice enthalpy (as in this case), the net effect is that the

overall change becomes more endothermic (or less exothermic in other possible cases where the total enthalpy change turns out to be negative).

2. If the lattice enthalpy falls faster than the hydration enthalpy, the opposite happens - the change will become less endothermic (or more exothermic).

It turns out that the main factor is the size of the negative ion.

1. For small negative ions like hydroxide, the lattice enthalpy falls faster than the hydration enthalpy of the positive ions.

That means that the enthalpy of solution will become less positive (or more negative).

2. For large negative ions like sulphate or carbonate, the hydration enthalpy of the positive ions falls faster than the lattice enthalpy. In this case, the enthalpy of solution will become more positive (or less negative).

With sulphates, for example, the percentage increase in the inter-ionic distance as you go from magnesium to calcium sulphate isn't as great as it would be with a smaller negative ion like hydroxide.Since the percentage increase in inter-ionic distance isn't very great, the change in the lattice enthalpy won't be very great either.

The Relationship Between Enthalpy of Solution and solubility

The assumption is made that the more endothermic (or less exothermic) the enthalpy of solution is, the less soluble the compound.So sulphates and carbonates become less soluble as you go down the Group; hydroxides become more soluble. The enthalpy of solution values for the Group 1 chlorides.

	Calculated value (kJ mol^{-1})	Data book value(kJ mol^{-1})
NaCl	+1	+3.9
KCl	+15	+17.2
RbCl	+10	+16.7
CsCl	+5	+17.9

The reasons for the discrepancies lie in the way the numbers are calculated. These are very small numbers worked out from much larger ones. Small uncertainties in those large numbers will cause large swings in the answers.The solubilities of the Group 1 chlorides (in moles of solute saturating 100 g of water at 298 K) compared with their enthalpies of solution are:

	Calculated value (kJ mol^{-1})	Solubility (mol/100 g of water
NaCl	+3.9	0.615
KCl	+17.2	0.481
RbCl	+16.7	0.781
CsCl	+17.9	1.13

There is no obvious relationship connecting the relative movements of these solubility values with the enthalpy of solution figures. It would be quite untrue to say that the more endothermic the change, the less soluble the compound.The table above illustrates this problem.Taking the sign of enthalpy of solution at face value.

$MgCO_3$	-25.3
$CaCO_3$	-12.3
$SrCO_3$	-3.4
$BaCO_3$	+4.2

Sodium chloride and the other Group 1 chlorides dissolve despite the fact that their enthalpies of solution are positive, and yet magnesium carbonate (and most of the other Group 2 carbonates) are very sparingly soluble, but have exothermic enthalpies of solution.The Group 1 chlorides, you could argue that the figures get progressively less exothermic, and at barium carbonate become endothermic. That would seem to support the decrease in solubility as you go down the Group quite nicely.If you look at the solubility data, the trend is broken at the bottom of the Group. Barium carbonate is more soluble than strontium carbonate.

To get around the problem of many compounds dissolving freely in water despite the fact that their enthalpies of solution are endothermic you have to introduce the concept of *entropy change*.The only way of making sense of entropy without getting bogged down in some serious maths is to think of it as a measure of the amount of disorder in a system. Entropy is given the symbol S. If a system becomes more disordered, then its entropy increases.In order to see whether a change is possible or not, you have to think about a combination of the enthalpy change and the entropy change. These can be combined mathematically to give an important term known as *free energy change*.

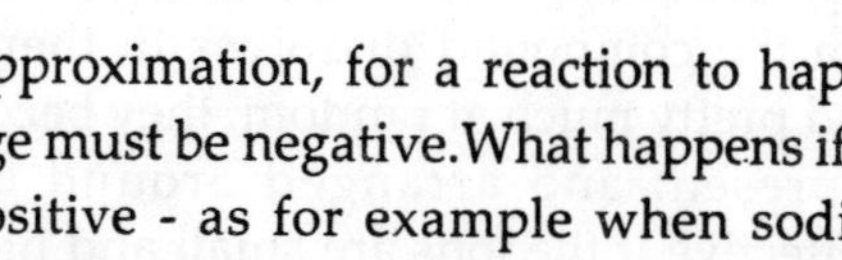

As an approximation, for a reaction to happen, the free energy change must be negative.What happens if the enthalpy change is positive - as for example when sodium chloride dissolves in water.As long as the entropy change is positive enough, it is possible to get a negative value for free energy change. In the sodium chloride case, you don't have to have very much increase in entropy to outweigh the small enthalpy change.

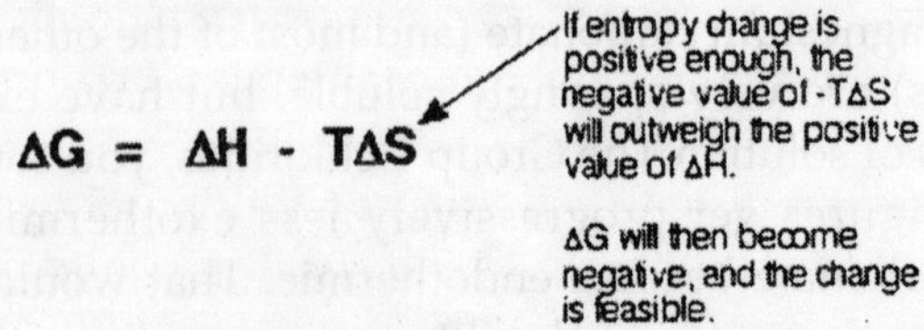

So . . . does the entropy increase when sodium chloride dissolve in water? Yes, it does! The sodium and chloride ions were arranged in a very tidy way in the crystal lattice - their entropy was low.When you dissolve the crystal in water, the entropy increases as the ions and water molecules become completely jumbled up - they become much more disordered than they were originally.

The most obvious thing that's wrong is that it won't explain why some compounds (like magnesium carbonate, and most of the other Group 2 carbonates) don't dissolve in water even though their enthalpies of solution are mainly negative.In these cases, the entropy of the system must fall when the compounds dissolve in water - in other words, the solution in water is more ordered than the original crystal and water!

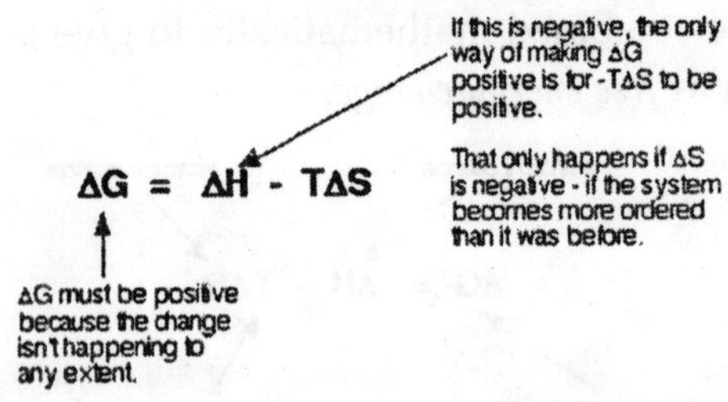

This happens because the water molecules become more ordered when the compound dissolves in them. Instead of milling around pretty much at random, they become attracted to the ions present and arranged around them.This is particularly effective if the ions are small and highly charged - and so the effect is greatest for the positive ions at the top of the Group, and gets less as you go down. It is also much more important in Group 2 than in Group 1 where the ions only carry one positive charge.That means that you have two entropy effects to consider. There is the increase in disorder as the crystal lattice breaks up, but a corresponding increase in order in the water - which varies depending on the sizes and charges of the ions present. To explain this properly, you need to think about the way lattice enthalpy changes as you go down the Group, the way that hydration enthalpies change, and the way that entropy changes. The way those changes happen will vary from one type of compound to another.

Remember that the solubility of the carbonates falls as you go down Group 2, apart from an increase as you go from strontium to barium carbonate.The general fall is because hydration enthalpies are falling faster than lattice enthalpies. Remember that where you have a big negative ion, its size dominates the inter-ionic distance and so doesn't allow the lattice enthalpy to change much. This gives the enthalpy of solution values we've already looked at (values in kJ mol^{-1}):

$MgCO_3$	-25.3
$CaCO_3$	-12.3
$SrCO_3$	-3.4
$BaCO_3$	+4.2

But the entropy change will also be varying as you go down the Group. At the top, where you have small 2+ ions, the overall entropy change in the system must be negative - the system as a whole becomes more ordered when the compound dissolves because of the way the water molecules become organised around the positive ions. That negative entropy change is going to be enough to wipe out the effect of the exothermic enthalpy of solution.

Towards the bottom of the Group, this effect changes. The bigger ions have less organising effect on the water molecules. The entropy change is becoming less negative (or perhaps even at this stage, positive). That's going to tend to make the compounds more soluble.The overall effect is a complex balance between the way the enthalpy of solution varies and the way the entropy change of solution alters. At barium carbonate, the effect of increasing entropy must be enough to make it more soluble than strontium carbonate.

THERMAL STABILITY

The effect of heat on the carbonates and nitrates of the Group 2 elements - beryllium, magnesium, calcium, strontium and barium. It describes and explains how the thermal stability of the compounds changes as you go down the Group.

All the carbonates in this Group undergo thermal decomposition to give the metal oxide and carbon dioxide. Thermal decomposition is the term given to splitting up a compound by heating it.If "X" represents any one of the elements:

$$XCO_{3(s)} \longrightarrow XO_{(s)} + CO_{2(g)}$$

As you go down the Group, the carbonates have to be heated more strongly before they will decompose.

The carbonates become more stable to heat as you go down the Group.

All the nitrates in this Group undergo thermal decomposition to give the metal oxide, nitrogen dioxide and oxygen.Again, if "X" represents any one of the elements:

$$2X(NO_3)_{2(s)} \longrightarrow 2XO_{(s)} + 4NO_{2(g)} + O_{2(g)}$$

As you go down the Group, the nitrates also have to be heated more strongly before they will decompose.

The nitrates also become more stable to heat as you go down the Group.

Both carbonates and nitrates become more thermally stable as you go down the Group. The ones lower down have to be heated more strongly than those at the top before they will decompose.A small 2+ ion has a lot of charge packed into a small volume of space. It has a high charge density and will have a marked distorting effect on any negative ions which happen to be near it.A bigger 2+ ion has the same charge spread over a larger volume of space. Its charge density will be lower, and it will cause less distortion to nearby negative ions.

Carbonate Ion

If you worked out the structure of a carbonate ion using "dots-and-crosses" or some similar method, you would probably come up with:

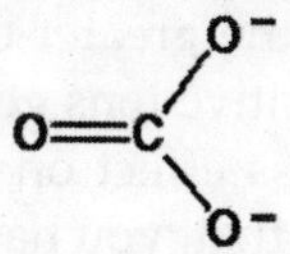

This shows two single carbon-oxygen bonds and one double one, with two of the oxygens each carrying a negative charge. Unfortunately, in real carbonate ions all the bonds are identical, and the charges are spread out over the whole ion - although concentrated on the oxygen atoms. We say that the charges are delocalised.This is a rather more complicated version of the bonding you might have come across in benzene or in ions like ethanoate.

The next diagram shows the delocalised electrons. The shading is intended to show that there is a greater chance of finding them around the oxygen atoms than near the carbon.

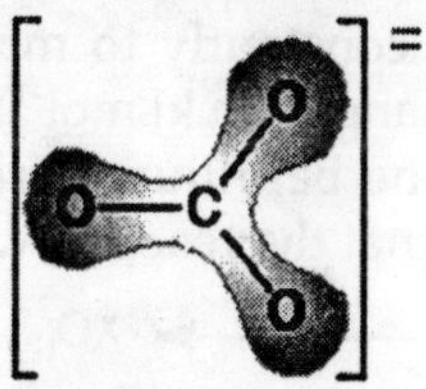

Polarising the carbonate ion

Now imagine what happens when this ion is placed next to a positive ion. The positive ion attracts the delocalised electrons in the carbonate ion towards itself. The carbonate ion becomes polarised.

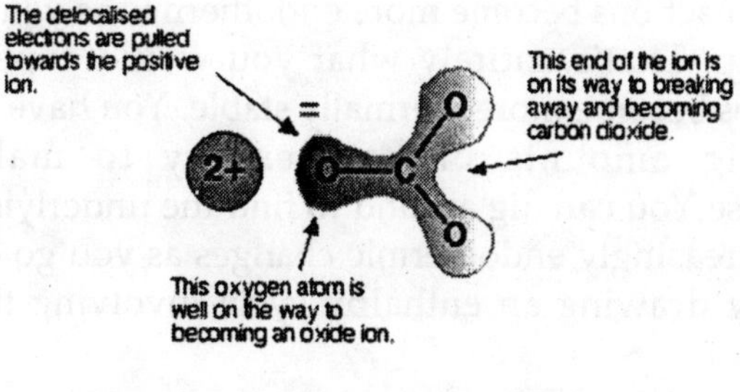

If this is heated, the carbon dioxide breaks free to leave the metal oxide.How much you need to heat the carbonate before that happens depends on how polarised the ion was. if it is highly polarised, you need less heat than if it is only slightly polarised.The smaller the positive ion is, the higher the charge density, and the greater effect it will have on the carbonate ion. As the positive ions get bigger as you go down the Group, they have less effect on the carbonate ions near them. To compensate for that, you have to heat the compound more in order to persuade the carbon dioxide to break free and leave the metal oxide.In other words, as you go down the Group, the carbonates become more thermally stable.

The argument is exactly the same here. The small positive ions at the top of the Group polarise the nitrate ions more than the larger positive ions at the bottom. Drawing diagrams to show this happening is much more difficult because the process has interactions involving more than one nitrate ion. If you calculate the enthalpy changes for the decompostion of the various carbonates, you find that all the changes are quite strongly endothermic. That implies that the reactions are likely to have to be heated constantly to make them happen.The calculated enthalpy changes (in kJ mol^{-1}) are given in the table. Figures to calculate the beryllium carbonate value weren't available. Remember that the reaction we are talking about is:

$$XCO_{3(s)} \longrightarrow XO_{(s)} + CO_{2(g)}$$

$MgCO_3$	+117
$CaCO_3$	+178
$SrCO_3$	+235
$BaCO_3$	+267

The reactions become more endothermic as you go down the Group. That's entirely what you would expect as the carbonates become more thermally stable. You have to supply increasing amounts of heat energy to make them decompose.You can dig around to find the underlying causes of the increasingly endothermic changes as you go down the Group by drawing an enthalpy cycle involving the lattice enthalpies of the metal carbonates and the metal oxides.There are two ways of defining lattice enthalpy. In order to make the argument mathematically simpler.*Lattice enthalpy* is the heat needed to split one mole of crystal in its standard state into its separate gaseous ions. For example, for magnesium oxide, it is the heat needed to carry out 1 mole of this change:

$$MgO_{(s)} \longrightarrow Mg^{2+}_{(g)} + O^{2-}_{(g)} \qquad LE = +3889\ kJ\ mol^{-1}$$

The cycle we are interested in looks like this:

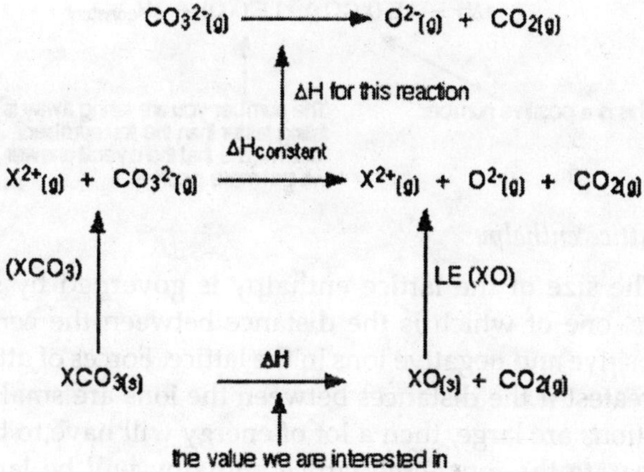

You can apply Hess's Law to this, and find two routes which will have an equal enthalpy change because they start and end in the same places.

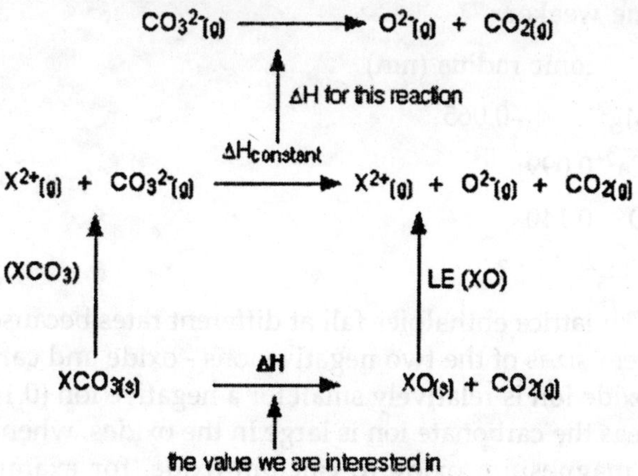

For reasons we will look at shortly, the lattice enthalpies of both the oxides and carbonates fall as you go down the Group. But they don't fall at the same rate.The oxide lattice enthalpy falls faster than the carbonate one. If you think carefully about what happens to the value of the overall enthalpy change of the decomposition reaction, you will see that it gradually becomes more positive as you go down the Group.

$$\Delta H = LE(XCO_3) - LE(XO) + \Delta H_{constant}$$

This is a positive number.

The number you are taking away is falling faster than the first number. That means that the overall answer will get more positive.

lattice enthalpy

The size of the lattice enthalpy is governed by several factors, one of which is the distance between the centres of the positive and negative ions in the lattice. Forces of attraction are greatest if the distances between the ions are small. If the attractions are large, then a lot of energy will have to be used to separate the ions - the lattice enthalpy will be large.The lattice enthalpies of both carbonates and oxides fall as you go down the Group because the positive ions are getting bigger. The inter-ionic distances are increasing and so the attractions become weaker.

	ionic radius (nm)
Mg^{2+}	0.065
Ca^{2+}	0.099
O^{2-}	0.140
CO_3^{2-}	?

The lattice enthalpies fall at different rates because of the different sizes of the two negative ions - oxide and carbonate. The oxide ion is relatively small for a negative ion (0.140 nm), whereas the carbonate ion is large.In the oxides, when you go from magnesium oxide to calcium oxide, for example, the inter-ionic distance increases from 0.205 nm (0.140 + 0.065) to 0.239 nm (0.140 + 0.099) - an increase of about 17%.In the carbonates, the inter-ionic distance is dominated by the much larger carbonate ion. Although the inter-ionic distance will increase by the same amount as you go from magnesium carbonate to calcium carbonate, as a percentage of the total distance the increase will be much less.The carbonate ion radius was 0.3 nm. The inter-ionic distances in the two cases we are talking about would increase from 0.365 nm to 0.399 nm - an increase of only about 9%.The rates at which the two lattice energies fall as you go down the Group depends on the percentage change as you go from one compound to the next. On that basis, the oxide lattice enthalpies are bound to fall faster than those of the carbonates.

Nitrates

The nitrate ion is bigger than an oxide ion, and so its radius tends to dominate the inter-ionic distance. The lattice enthalpy of the oxide will again fall faster than the nitrate.

Beryllium Chloride is Covalent

Beryllium chloride, $BeCl_2$, melts at 405°C and boils at 520°C. That compares with 714°C and 1412°C for magnesium chloride.Notice how much dramatically lower the boiling point of beryllium chloride is compared with magnesium chloride. The much higher boiling point of magnesium chloride is what you might expect from the strong forces between the positive and negative ions present.Because its boiling point is much lower, it follows that beryllium chloride can't contain ions - it must be covalent. On the other hand, the melting point is quite high for such a small covalent molecule.

Reaction with Water

Beryllium chloride reacts vigorously and exothermically with water with the evolution of acidic, steamy hydrogen chloride gas. This is typical of covalent chlorides.In the first instance, it reacts to give hydrated beryllium ions, $[Be(H_2O)_4]^{2+}$, and chloride ions.But the hydrated beryllium ions (called tetraaquaberyllium ions) are quite strongly acidic.

The small beryllium ion at the centre attracts the electrons in the bonds towards itself, and that makes the hydrogen atoms in the water even more positive than they usually are. If the solution is hot and concentrated (as it is likely to be if you add water to solid beryllium chloride - a very exothermic reaction), chloride ions can remove one or more of these hydrogen ions to produce hydrogen chloride gas.All the other ionic chlorides in Group 2 dissolve in water without any obvious reaction.

The Structure of Beryllium Chloride

Beryllium chloride, $BeCl_2$, is a linear molecule with all three atoms in a straight line. Showing only the outer electrons:

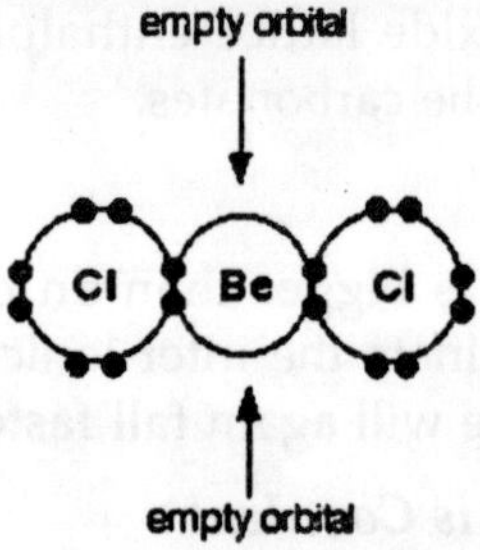

Beryllium chloride is known as an *electron-deficient compound* because it has the two empty orbitals at the bonding level.If it had this same simple structure as a solid, you would expect the melting point to be much lower than it actually is. It is a very small molecule, and so the intermolecular attractions would be expected to be fairly weak.In the solid, the $BeCl_2$ molecules polymerise to make long chains. They do this by forming coordinate bonds (dative covalent bonds) between lone pairs on chlorine atoms and adjacent beryllium atoms. The diagram shows a simple dimer - the start of the polymerisation process.

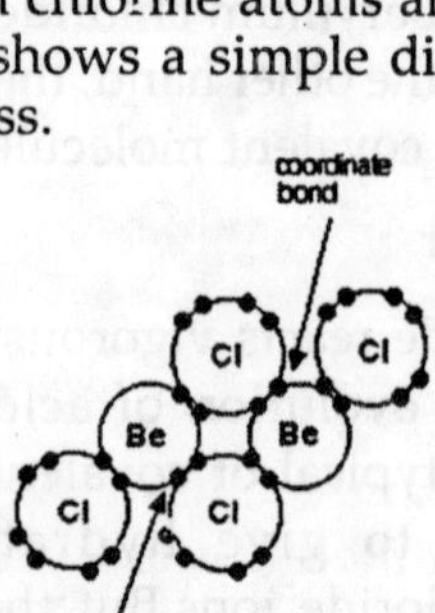

The last diagram that the beryllium atoms are still electron deficient. The process continues. The next diagram shows the coordinate bonds in the conventional way using arrows. The arrow goes from the atom which is supplying the pair of electrons to the atom with the empty orbital.

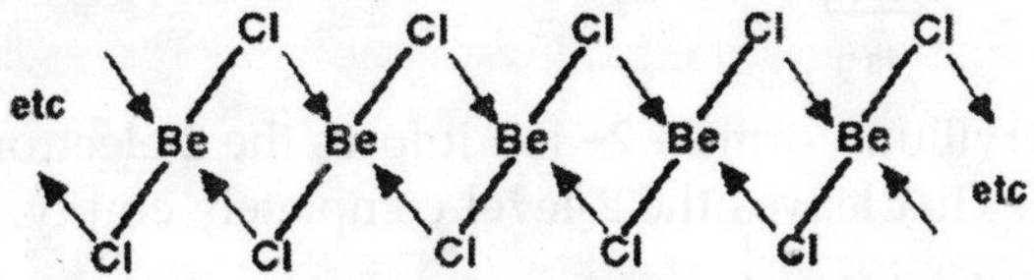

Make sure that you fully understand how this diagram relates to the dimer shown in the previous diagram.Beryllium has quite a high electronegativity compared with the rest of the Group. That means that it attracts a bonding pair of electrons towards itself more strongly than magnesium and the rest do.In order for an ionic bond to form, the beryllium has to let go of its electrons. It is too electronegative to do that.

Beryllium Forms 4-coordinated Complex Ions

Although beryllium doesn't normally form simple ions, Be^{2+}, it does form ions in solution. In these, the beryllium ion becomes attached to four water molecules to give a complex ion with the formula $[Be(H_2O)_4]^{2+}$.The ion is said to be 4-*coordinated,* or to have a *coordination number of 4,* because there are four water molecules arranged around the central beryllium.Many hydrated metal ions are 6-coordinated. For example, magnesium ions in solution exist as $[Mg(H_2O)_6]^{2+}$.The water molecules in these ions are attached to the central metal ion via coordinate bonds (dative covalent bonds). One of the lone pairs on each water molecule is used to form a bond with an empty orbital in the metal ion.Each time one of these bonds is formed, energy is released, and the ion becomes more stable.

The Hydration of Beryllium

The problem is that there has to be somewhere that the lone pairs on the water molecules can attach to. Beryllium has the electronic structure $1s^22s^2$. It is helpful to draw this as an "electrons-in-boxes" diagram:

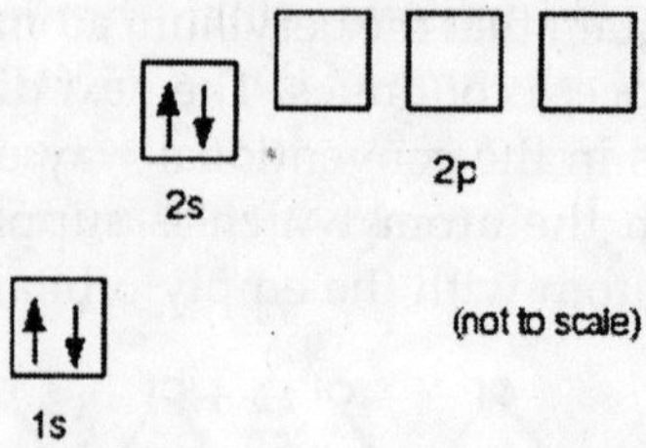

When beryllium forms a 2+ ion it loses the 2 electrons in the 2s orbital. That leaves the 2-level completely empty.

The 2-level orbitals reorganise themselves (hybridise) to make four equal orbitals, each of which can accept a lone pair of electrons from a water molecule. In the next diagram the 1s electrons have been left out. They aren't relevant to the bonding.

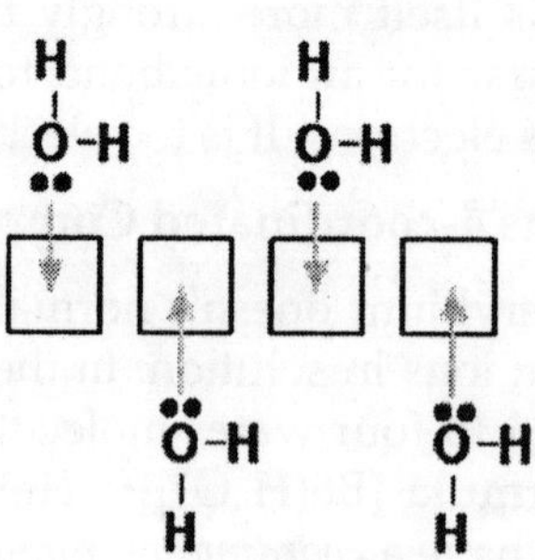

Each water molecule, of course, has two lone pairs of electrons. Only one of them is shown to avoid cluttering the diagram.

Notice that once four water molecules have bonded in this way, there isn't any more space available at the bonding level. All the empty orbitals from the original beryllium ion are being used.The water molecules arrange themselves to get as far apart as possible - which is pointing towards the corners of a tetrahedron. The ion therefore has a tetrahedral shape.

The hydration of magnesium

You might think that magnesium would behave just the same, but at the 3-level there are 3d orbitals available as well as 3s and 3p.When the magnesium ion is formed, it leaves empty 3s, 3p and 3d orbitals. When that ion is hydrated, it uses the 3s orbital, all three of the 3p orbitals and two of the 3d orbitals. These are reorganised to leave a total of six empty orbitals which are then used for bonding.

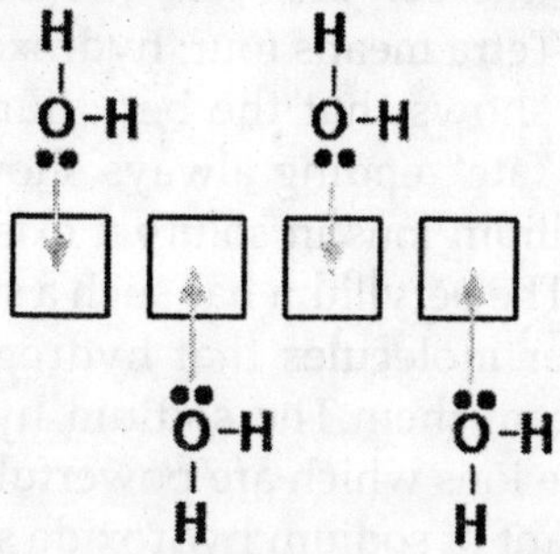

As the ions get bigger, there is less tendency for them to form proper coordinate bonds with water molecules. The ions become so big that they aren't sufficiently attractive to the lone pairs on the water molecules to form formal bonds - instead the water molecules tend to cluster more loosely around the positive ions.Where they do form coordinate bonds with the water, however, they will be 6-coordinated just like the magnesium.

Beryllium hydroxide is amphoteric

Amphoteric means that it can react with both acids and bases to form salts.

The other hydroxides of the Group 2 metals are all basic. They react with acids to form salts. For example:

$$Ca(OH)_{2(s)} + 2HCl_{(aq)} \longrightarrow CaCl_{2(aq)} + 2H_2O_{(l)}$$

Calcium hydroxide reacts with dilute hydrochloric acid to give calcium chloride and water.

Beryllium hydroxide

Beryllium hydroxide reacts with acids, forming solutions of beryllium salts. For example:

$$Be(OH)_{2(s)} + H_2SO_{4(aq)} \longrightarrow BeSO_{4(aq)} + 2H_2O_{(l)}$$

But it also reacts with bases such as sodium hydroxide solution. Beryllium hydroxide reacts with the sodium hydroxide to give a colourless solution of sodium tetrahydroxoberyllate.

$$Be(OH)_{2(s)} + 2NaOH_{(aq)} \longrightarrow Na_2Be(OH)_{4(aq)}$$

This contains the complex ion, $[Be(OH)_4]^{2-}$. The name describes this ion. Tetra means four; hydroxo refers to the OH groups; beryllate shows that the beryllium is present in a negative ion. The "ate" ending always shows that the ion is negative.The beryllium ions in solut• n exist as the hydrated ion, $[Be(H_2O)_4]^{2+}$. The beryllium has such a strongly polarising effect on the water molecules that hydrogen ions are very easily removed from them.The sodium hydroxide solution contains hydroxide ions which are powerful bases. If you add just the right amount of sodium hydroxide solution, you get a precipitate of what is normally called "beryllium hydroxide" - but which is a shade more complicated than that!

$$[Be(H_2O)_4]^{2+}{}_{(aq)} + 2OH^-{}_{(aq)} \longrightarrow Be(H_2O)_2(OH)_{2(s)} + 2H_2O_{(l)}$$

The product (other than water) is a neutral complex, and it is covalently bonded. All that has happened to the original complex ion is that two hydrogen ions have been removed from the water molecules.You get a precipitate of the neutral complex because of the lack of charge on it. There isn't enough attraction between this neutral complex and water molecules to bring it into solution.

The hydrogen ions that were originally removed are simply replaced. The precipitate dissolves as the original hydrated beryllium ion is re-formed.

$$Be(H_2O)_2(OH)_{2(s)} + 2H^+{}_{(aq)} \longrightarrow [Be(H_2O)_4]^{2+}{}_{(aq)}$$

Adding more hydroxide ions to the neutral complex pulls more hydrogen ions off the water molecules to give the tetrahydroxoberyllate ion:

$$Be(H_2O)_2(OH)_{2(s)} + 2OH^-{}_{(aq)} \longrightarrow [Be(OH)_4]^{2-}{}_{(aq)} + 2H_2O_{(l)}$$

The beryllium hydroxide dissolves because the neutral complex is converted into an ion which will be sufficiently attracted to water molecules.

Calcium hydroxide is truly ionic - and contains simple hydroxide ions, OH^-. These react with hydrogen ions from an acid to form water - and so the hydroxide reacts with acids. There isn't any equivalent to the neutral complex. Adding more hydroxide ions from a base has no effect because they haven't got anything to react with.

